Statistical Methods for Healthcare Performance Monitoring

Chapman & Hall/CRC Biostatistics Series

Editor-in-Chief

Shein-Chung Chow, Ph.D., Professor, Department of Biostatistics and Bioinformatics,
Duke University School of Medicine, Durham, North Carolina

Series Editors

Byron Jones, Biometrical Fellow, Statistical Methodology, Integrated Information Sciences,
Novartis Pharma AG, Basel, Switzerland

Jen-pei Liu, Professor, Division of Biometry, Department of Agronomy,
National Taiwan University, Taipei, Taiwan

Karl E. Peace, Georgia Cancer Coalition, Distinguished Cancer Scholar, Senior Research Scientist
and Professor of Biostatistics, Jiann-Ping Hsu College of Public Health,
Georgia Southern University, Statesboro, Georgia

Bruce W. Turnbull, Professor, School of Operations Research and Industrial Engineering,
Cornell University, Ithaca, New York

Published Titles

Adaptive Design Methods in Clinical Trials, Second Edition
Shein-Chung Chow and Mark Chang

Adaptive Designs for Sequential Treatment Allocation
Alessandro Baldi Antognini
and Alessandra Giovagnoli

Adaptive Design Theory and Implementation Using SAS and R, Second Edition
Mark Chang

Advanced Bayesian Methods for Medical Test Accuracy
Lyle D. Broemeling

Applied Biclustering Methods for Big and High-Dimensional Data Using R
Adetayo Kasim, Ziv Shkedy,
Sebastian Kaiser, Sepp Hochreiter,
and Willem Talloen

Applied Meta-Analysis with R
Ding-Geng (Din) Chen and Karl E. Peace

Basic Statistics and Pharmaceutical Statistical Applications, Second Edition
James E. De Muth

Bayesian Adaptive Methods for Clinical Trials
Scott M. Berry, Bradley P. Carlin,
J. Jack Lee, and Peter Muller

Bayesian Analysis Made Simple: An Excel GUI for WinBUGS
Phil Woodward

Bayesian Designs for Phase I–II Clinical Trials
Ying Yuan, Hoang Q. Nguyen,
and Peter F. Thall

Bayesian Methods for Measures of Agreement
Lyle D. Broemeling

Bayesian Methods for Repeated Measures
Lyle D. Broemeling

Bayesian Methods in Epidemiology
Lyle D. Broemeling

Bayesian Methods in Health Economics
Gianluca Baio

Bayesian Missing Data Problems: EM, Data Augmentation and Noniterative Computation
Ming T. Tan, Guo-Liang Tian,
and Kai Wang Ng

Bayesian Modeling in Bioinformatics
Dipak K. Dey, Samiran Ghosh,
and Bani K. Mallick

Benefit-Risk Assessment in Pharmaceutical Research and Development
Andreas Sashegyi, James Felli,
and Rebecca Noel

Published Titles

Benefit-Risk Assessment Methods in Medical Product Development: Bridging Qualitative and Quantitative Assessments
Qi Jiang and Weili He

Biosimilars: Design and Analysis of Follow-on Biologics
Shein-Chung Chow

Biostatistics: A Computing Approach
Stewart J. Anderson

Cancer Clinical Trials: Current and Controversial Issues in Design and Analysis
Stephen L. George, Xiaofei Wang, and Herbert Pang

Causal Analysis in Biomedicine and Epidemiology: Based on Minimal Sufficient Causation
Mikel Aickin

Clinical and Statistical Considerations in Personalized Medicine
Claudio Carini, Sandeep Menon, and Mark Chang

Clinical Trial Data Analysis using R
Ding-Geng (Din) Chen and Karl E. Peace

Clinical Trial Methodology
Karl E. Peace and Ding-Geng (Din) Chen

Computational Methods in Biomedical Research
Ravindra Khattree and Dayanand N. Naik

Computational Pharmacokinetics
Anders Källén

Confidence Intervals for Proportions and Related Measures of Effect Size
Robert G. Newcombe

Controversial Statistical Issues in Clinical Trials
Shein-Chung Chow

Data Analysis with Competing Risks and Intermediate States
Ronald B. Geskus

Data and Safety Monitoring Committees in Clinical Trials
Jay Herson

Design and Analysis of Animal Studies in Pharmaceutical Development
Shein-Chung Chow and Jen-pei Liu

Design and Analysis of Bioavailability and Bioequivalence Studies, Third Edition
Shein-Chung Chow and Jen-pei Liu

Design and Analysis of Bridging Studies
Jen-pei Liu, Shein-Chung Chow, and Chin-Fu Hsiao

Design & Analysis of Clinical Trials for Economic Evaluation & Reimbursement: An Applied Approach Using SAS & STATA
Iftekhar Khan

Design and Analysis of Clinical Trials for Predictive Medicine
Shigeyuki Matsui, Marc Buyse, and Richard Simon

Design and Analysis of Clinical Trials with Time-to-Event Endpoints
Karl E. Peace

Design and Analysis of Non-Inferiority Trials
Mark D. Rothmann, Brian L. Wiens, and Ivan S. F. Chan

Difference Equations with Public Health Applications
Lemuel A. Moyé and Asha Seth Kapadia

DNA Methylation Microarrays: Experimental Design and Statistical Analysis
Sun-Chong Wang and Arturas Petronis

DNA Microarrays and Related Genomics Techniques: Design, Analysis, and Interpretation of Experiments
David B. Allison, Grier P. Page, T. Mark Beasley, and Jode W. Edwards

Dose Finding by the Continual Reassessment Method
Ying Kuen Cheung

Dynamical Biostatistical Models
Daniel Commenges and Hélène Jacqmin-Gadda

Elementary Bayesian Biostatistics
Lemuel A. Moyé

Empirical Likelihood Method in Survival Analysis
Mai Zhou

Exposure–Response Modeling: Methods and Practical Implementation
Jixian Wang

Published Titles

Frailty Models in Survival Analysis
Andreas Wienke

Fundamental Concepts for New Clinical Trialists
Scott Evans and Naitee Ting

Generalized Linear Models: A Bayesian Perspective
Dipak K. Dey, Sujit K. Ghosh, and Bani K. Mallick

Handbook of Regression and Modeling: Applications for the Clinical and Pharmaceutical Industries
Daryl S. Paulson

Inference Principles for Biostatisticians
Ian C. Marschner

Interval-Censored Time-to-Event Data: Methods and Applications
Ding-Geng (Din) Chen, Jianguo Sun, and Karl E. Peace

Introductory Adaptive Trial Designs: A Practical Guide with R
Mark Chang

Joint Models for Longitudinal and Time-to-Event Data: With Applications in R
Dimitris Rizopoulos

Measures of Interobserver Agreement and Reliability, Second Edition
Mohamed M. Shoukri

Medical Biostatistics, Third Edition
A. Indrayan

Meta-Analysis in Medicine and Health Policy
Dalene Stangl and Donald A. Berry

Mixed Effects Models for the Population Approach: Models, Tasks, Methods and Tools
Marc Lavielle

Modeling to Inform Infectious Disease Control
Niels G. Becker

Modern Adaptive Randomized Clinical Trials: Statistical and Practical Aspects
Oleksandr Sverdlov

Monte Carlo Simulation for the Pharmaceutical Industry: Concepts, Algorithms, and Case Studies
Mark Chang

Multiregional Clinical Trials for Simultaneous Global New Drug Development
Joshua Chen and Hui Quan

Multiple Testing Problems in Pharmaceutical Statistics
Alex Dmitrienko, Ajit C. Tamhane, and Frank Bretz

Noninferiority Testing in Clinical Trials: Issues and Challenges
Tie-Hua Ng

Optimal Design for Nonlinear Response Models
Valerii V. Fedorov and Sergei L. Leonov

Patient-Reported Outcomes: Measurement, Implementation and Interpretation
Joseph C. Cappelleri, Kelly H. Zou, Andrew G. Bushmakin, Jose Ma. J. Alvir, Demissie Alemayehu, and Tara Symonds

Quantitative Evaluation of Safety in Drug Development: Design, Analysis and Reporting
Qi Jiang and H. Amy Xia

Quantitative Methods for Traditional Chinese Medicine Development
Shein-Chung Chow

Randomized Clinical Trials of Nonpharmacological Treatments
Isabelle Boutron, Philippe Ravaud, and David Moher

Randomized Phase II Cancer Clinical Trials
Sin-Ho Jung

Sample Size Calculations for Clustered and Longitudinal Outcomes in Clinical Research
Chul Ahn, Moonseong Heo, and Song Zhang

Published Titles

Sample Size Calculations in Clinical Research, Second Edition
Shein-Chung Chow, Jun Shao, and Hansheng Wang

Statistical Analysis of Human Growth and Development
Yin Bun Cheung

Statistical Design and Analysis of Clinical Trials: Principles and Methods
Weichung Joe Shih and Joseph Aisner

Statistical Design and Analysis of Stability Studies
Shein-Chung Chow

Statistical Evaluation of Diagnostic Performance: Topics in ROC Analysis
Kelly H. Zou, Aiyi Liu, Andriy Bandos, Lucila Ohno-Machado, and Howard Rockette

Statistical Methods for Clinical Trials
Mark X. Norleans

Statistical Methods for Drug Safety
Robert D. Gibbons and Anup K. Amatya

Statistical Methods for Healthcare Performance Monitoring
Alex Bottle and Paul Aylin

Statistical Methods for Immunogenicity Assessment
Harry Yang, Jianchun Zhang, Binbing Yu, and Wei Zhao

Statistical Methods in Drug Combination Studies
Wei Zhao and Harry Yang

Statistical Testing Strategies in the Health Sciences
Albert Vexler, Alan D. Hutson, and Xiwei Chen

Statistics in Drug Research: Methodologies and Recent Developments
Shein-Chung Chow and Jun Shao

Statistics in the Pharmaceutical Industry, Third Edition
Ralph Buncher and Jia-Yeong Tsay

Survival Analysis in Medicine and Genetics
Jialiang Li and Shuangge Ma

Theory of Drug Development
Eric B. Holmgren

Translational Medicine: Strategies and Statistical Methods
Dennis Cosmatos and Shein-Chung Chow

Chapman & Hall/CRC Biostatistics Series

Statistical Methods for Healthcare Performance Monitoring

Alex Bottle
Imperial College London, United Kingdom

Paul Aylin
Imperial College London, United Kingdom

CRC Press
Taylor & Francis Group
Boca Raton   London   New York

CRC Press is an imprint of the
Taylor & Francis Group, an **informa** business

A CHAPMAN & HALL BOOK

CRC Press
Taylor & Francis Group
6000 Broken Sound Parkway NW, Suite 300
Boca Raton, FL 33487-2742

© 2017 by Taylor & Francis Group, LLC
CRC Press is an imprint of Taylor & Francis Group, an Informa business

No claim to original U.S. Government works

ISBN-13: 978-1-4822-4609-4 (hbk)

This book contains information obtained from authentic and highly regarded sources. Reasonable efforts have been made to publish reliable data and information, but the author and publisher cannot assume responsibility for the validity of all materials or the consequences of their use. The authors and publishers have attempted to trace the copyright holders of all material reproduced in this publication and apologize to copyright holders if permission to publish in this form has not been obtained. If any copyright material has not been acknowledged please write and let us know so we may rectify in any future reprint.

Except as permitted under U.S. Copyright Law, no part of this book may be reprinted, reproduced, transmitted, or utilized in any form by any electronic, mechanical, or other means, now known or hereafter invented, including photocopying, microfilming, and recording, or in any information storage or retrieval system, without written permission from the publishers.

For permission to photocopy or use material electronically from this work, please access www.copyright.com (http://www.copyright.com/) or contact the Copyright Clearance Center, Inc. (CCC), 222 Rosewood Drive, Danvers, MA 01923, 978-750-8400. CCC is a not-for-profit organization that provides licenses and registration for a variety of users. For organizations that have been granted a photocopy license by the CCC, a separate system of payment has been arranged.

Trademark Notice: Product or corporate names may be trademarks or registered trademarks, and are used only for identification and explanation without intent to infringe.

Library of Congress Cataloging-in-Publication Data

Names: Bottle, Alex, author. | Aylin, Paul, author.
Title: Statistical methods for healthcare performance monitoring / Alex Bottle and Paul Aylin.
Description: Boca Raton : Taylor & Francis, 2017. | Series: Chapman & Hall/CRC biostatistics series ; 92 | "A CRC title, part of the Taylor & Francis imprint, a member of the Taylor & Francis Group, the academic division of T&F Informa plc." | Includes bibliographical references and index.
Identifiers: LCCN 2016006496 | ISBN 9781482246094 (alk. paper)
Subjects: LCSH: Medical care--Evaluation. | Medical statistics. | Medical care--Quality control. | Medical care--Safety measures.
Classification: LCC RA399.A1 B68 2016 | DDC 362.1072/7--dc23
LC record available at https://lccn.loc.gov/2016006496

Visit the Taylor & Francis Web site at
http://www.taylorandfrancis.com

and the CRC Press Web site at
http://www.crcpress.com

Contents

List of Illustrations xv
List of Tables xvii
Preface xix
Authors xxi

1. Introduction 1
   1.1   The Need for Performance Monitoring 1
   1.2   Measuring and Monitoring Quality 2
   1.3   The Need for This Book 2
   1.4   Who Is This Book For and How Should It Be Used? 3
   Common Abbreviations Used in the Book 4
   Acknowledgment 6

2. Origins and Examples of Monitoring Systems 7
   Aims of This Chapter 7
   2.1   Origins 7
   2.2   Healthcare Scandals 8
         2.2.1   Responses to the Scandals 12
   2.3   Examples of Monitoring Schemes 14
   2.4   Goals of Monitoring 15
         2.4.1   Accountability 15
         2.4.2   Regulation and Accreditation 16
         2.4.3   Patient Choice 19
         2.4.4   Openness and Transparency 21
         2.4.5   Quality Improvement 21
         2.4.6   Prevent Harm and Unsafe Care 22
         2.4.7   Professionalism 23
         2.4.8   Informed Consent 23

3. Choosing the Unit of Analysis and Reporting 25
   Aims of This Chapter 25
   3.1   Issues Principally Concerning the Analysis 26
         3.1.1   Clustering (*) 26
         3.1.2   Episode Treatment Groups 32
   3.2   Issues More Relevant to Reporting: Attributing Performance
         to a Given Unit in a System 34

4. What to Measure: Choosing and Defining Indicators ... 37
Aims of This Chapter ... 37
- 4.1 How Can We Define Quality? ... 38
- 4.2 Common Indicator Taxonomies ... 38
- 4.3 Particular Challenges of Measuring Patient Safety ... 42
- 4.4 Particular Challenges of Multimorbidity ... 45
- 4.5 Measuring the Health of the Population and Quality of the Whole Healthcare System ... 47
  - 4.5.1 The WHO Annual World Health Statistics Report ... 50
- 4.6 Efficiency and Value ... 53
  - 4.6.1 Data Envelopment Analysis and Stochastic Frontier Analysis (∗) ... 57
- 4.7 Features of an Ideal Indicator ... 59
- 4.8 Steps in Construction and Common Issues in Definition ... 59
- 4.9 Validation of Indicators ... 62
- 4.10 Some Strategies for Choosing among Candidates ... 63
- 4.11 Time to Go: When to Withdraw Indicators ... 64
- 4.12 Conclusion ... 65

5. Sources of Data ... 67
Aims of This Chapter ... 67
- 5.1 How to Assess Data Quality ... 67
- 5.2 Administrative Data ... 69
  - 5.2.1 Coding Systems for Administrative Data ... 71
  - 5.2.2 Use of Administrative Databases to Flag Patient Safety Events ... 74
- 5.3 Clinical Registry Data ... 75
- 5.4 Accuracy of Administrative and Clinical Databases Compared ... 77
- 5.5 Incident Reports and Other Ways to Capture Safety Events ... 84
- 5.6 Surveys ... 87
- 5.7 Other Sources ... 89
- 5.8 Other Issues Concerning Data Sources ... 90
- 5.9 Conclusion ... 91

6. Risk-Adjustment Principles and Methods ... 93
Aims of This Chapter ... 93
- 6.1 Risk Adjustment and Risk Prediction ... 94
- 6.2 When and Why Should We Adjust for Risk? ... 95
- 6.3 Alternatives to Risk Adjustment ... 96
- 6.4 What Factors Should We Adjust For? ... 96
  - 6.4.1 Factors Not under the Control of the Provider ... 96
  - 6.4.2 Proxies Such as Age and Socioeconomic Status ... 98
  - 6.4.3 Comorbidity ... 98
  - 6.4.4 Disease Severity ... 100

6.5	Selecting an Initial Set of Candidate Variables		101
6.6	Dealing with Missing and Extreme Values		102
6.7	Timing of the Risk Factor Measurement		104
6.8	Building the Model		108
	6.8.1	Choosing the Final Set of Variables from the Initial Set of Candidates	108
	6.8.2	Decide How Each Variable Should Be Entered into the Model	111
	6.8.3	Decide on the Statistical Method for Modelling (*)	111
	6.8.4	Assess the Fit of the Model (*)	114
		6.8.4.1 Adjusted R^2	115
		6.8.4.2 Area under the Receiver Operating Characteristic Curve: c Statistic	115
		6.8.4.3 The Hosmer–Lemeshow Statistic for Calibration	116
	6.8.5	Which Is More Important, Discrimination or Calibration?	117
	6.8.6	What Can Be Done If the Model Fit or Performance Is Unacceptable?	117
	6.8.7	Convert Regression Coefficients into a Risk Score If Desired	118

7. Output the Observed and Model-Predicted Outcomes (*) 121
Aims of This Chapter 121
7.1	Ratios versus Differences	122
7.2	Deriving SMRs from Standardisation and Logistic Regression	123
7.3	Other Fixed Effects Approaches to Generate an SMR	126
7.4	Random Effects–Based SMRs (*)	126
7.5	Marginal versus Multilevel Models (*)	129
7.6	Which Is the "Best" Modelling Approach Overall? (*)	130
7.7	Further Reading on Producing Risk-Adjusted Outcomes by Unit	134

8. Composite Measures 135
Aims of This Chapter 135
8.1	Some Examples		136
8.2	Steps in the Construction		137
	8.2.1	Specify the Scope and Purpose	138
	8.2.2	Choose the Unit	138
	8.2.3	Select the Data and Deal with Missing Values	138
	8.2.4	Choose the Indicators and Run Descriptive Analyses	138
	8.2.5	Normalise the Metrics	140
	8.2.6	Assign Weights and Aggregate the Component Indicators	140
	8.2.7	Run Sensitivity Analyses	144
	8.2.8	Present the Results	145

8.3		Some Real Examples 146
	8.3.1	AHRQ's Patient Safety Indicator Composite 146
	8.3.2	Leapfrog Group Patient Safety Composite 147
8.4		Pros and Cons of Composites.. 148
8.5		Alternatives to the Use of Composites.................................... 149

9. Setting Performance Thresholds and Defining Outliers.................. 151
Aims of This Chapter 151
- 9.1 Defining Acceptable Performance.. 152
  - 9.1.1 Targets 152
  - 9.1.2 Historical Benchmarks... 153
  - 9.1.3 Referring to Inter-Unit Variation 153
- 9.2 Bayesian Methods for Comparing Providers............................... 155
- 9.3 Statistical Process Control and Funnel Plots................................ 157
- 9.4 Multiple Testing (*).. 163
  - 9.4.1 Multivariate Statistical Process Control Methods (*)...... 165
  - 9.4.2 Further Reading on SPC 166
- 9.5 Ways of Assessing Variation between Units................................ 166
- 9.6 How Much Variation Is "Acceptable"?.. 167
- 9.7 Impact on Outlier Status of Using Fixed versus Random Effects to Derive SMRs 171
- 9.8 How Reliably Can We Detect Poor Performance?....................... 172
- 9.9 Some Resources for Quality Improvement Methods 174

10. Making Comparisons across National Borders 177
Aims of This Chapter 177
- 10.1 Examples of Multinational Patient-Level Databases................... 178
- 10.2 Challenges... 180
  - 10.2.1 Worked Example of Combining Administrative Databases from Multiple Countries: Stroke Mortality.........181
  - 10.2.2 Clustering within Countries 184
  - 10.2.3 Countries with Unusual Data or Apparent Performance 184
- 10.3 Interpreting Apparent Differences in Performance between Countries.. 185
- 10.4 Conclusion.. 186

11. Presenting the Results to Stakeholders 189
Aims of This Chapter 189
- 11.1 The Main Ways of Presenting Comparative Performance Data.... 189
- 11.2 Effect on Behaviour of the Choice of Format When Providing Performance Data.. 192
- 11.3 The Importance of the Method of Presentation........................... 195
  - 11.3.1 Presenting Performance Data to Managers and Clinicians... 195
  - 11.3.2 Presenting Results to the Public 197

11.4	Examples of Giving Performance Information to Units	198
11.5	Examples of Giving Performance Information to the Public	202
11.6	Metadata	206

12. Evaluating the Monitoring System ... 207
Aims of This Chapter ... 207
- 12.1 Study Design and Statistical Approaches to Evaluating a Monitoring System ... 208
  - 12.1.1 Interrupted Time-Series Design and Analysis (*) ... 209
  - 12.1.2 Adjusting for Confounding (*) ... 212
  - 12.1.3 Difference-in-Differences ... 214
  - 12.1.4 Instrumental Variable Analysis ... 216
  - 12.1.5 Regression Discontinuity Designs ... 217
- 12.2 Economic Evaluation Methods ... 220

13. Concluding Thoughts ... 225
- 13.1 Simple versus Complex ... 225
- 13.2 Specific versus General ... 226
- 13.3 The Future ... 227

Appendix A: Glossary of Main Statistical Terms Used ... 229

References ... 237

Index ... 261

List of Illustrations

Figure 6.1	Comparison of percentage of AVSD operations by age at operation (in months) between Bristol (UBHT) and elsewhere in England 1991/1992–1994/1995.	97
Figure 9.1	An example of a run chart showing a run of eight data points below the median.	158
Figure 9.2	An example of a Shewhart chart of the data from the run chart of Figure 9.1, with upper and lower control limits with 3 SD on either side of the mean.	159
Figure 9.3	An example of a funnel plot: inner lines represent 95% and outer lines 99.8% control limits.	160
Figure 11.1	Traffic light display of fictitious data on multiple indicators by unit.	190
Figure 11.2	Spider plot of an imaginary unit's fictitious data.	192
Figure 11.3	Benchmarked performance for NHS West Essex CCG's commissioned elective knee surgery.	200
Figure 11.4	Example scatter plot from Public Health England's commissioning tool "Fingertips" showing CCG-level deprivation and adult inactivity.	201
Figure 11.5	Screenshot from NHS Choices.	203
Figure 11.6	Illustrative Hospital Compare results for three Seattle hospitals for readmissions and deaths in patients with COPD.	205
Figure 11.7	COPD readmission rates for three Seattle hospitals displayed as bar charts by Hospital Compare.	205
Figure 12.1	Time segments pre- and post-intervention in an interrupted time series design.	211

List of Tables

Table 3.1	Main Methodological Issues When Considering the Unit of Analysis	29
Table 4.1	Structure, Process and Outcome Indicators Compared	40
Table 4.2	Ways of Combining Quality and Cost Measures in a 2014 National Quality Forum White Paper	56
Table 4.3	Attributes of Ideal Quality Indicators	60
Table 5.1	Comparison of Key Features of Clinical and Administrative Databases	83
Table 6.1	Issues Associated with Common Patient Factors in Risk Models	107
Table 7.1	Summary of the Main Types of *Standardised Mortality Ratio* Derived from Logistic Regression or Standardisation with the National Average as the Benchmark Unless Specified Otherwise	133
Table 8.1	Main Steps in Composite Indicator Construction	137
Table 8.2	Common Options for Weighting and Combining Component Indicators in a Composite	141
Table 8.3	Main Pros and Cons of Healthcare Composite Measures	148
Table 9.1	Common Ways to Compare Units	155
Table 9.2	Common Definitions of Outlying Performance	162
Table 10.1	Main Challenges for Maintaining an International Database for Benchmarking Outcomes	181
Table 11.1	Fictitious Data for Average Appointment Waiting Times by Unit with Categories and Star Rating	190
Table 11.2	Selected Public-Facing Websites for Obtaining Comparative Healthcare Unit Performance	203
Table 13.1	Trade-Offs in Healthcare Performance Monitoring	226

Preface

> Measurement is the first step that leads to control and eventually to improvement. If you can't measure something, you can't understand it. If you can't understand it, you can't control it. If you can't control it, you can't improve it.
>
> **H. James Harrington**

Healthcare is important to everyone. A high standard of healthcare is even more important, yet large variations in quality have been well documented both between and within many countries. As populations grow older and medical care becomes more complex, expenditures continue to rise. It is therefore more crucial than ever to know how well the healthcare system and all its components are performing. This requires data, which inevitably differ in form and quality, and it requires statistical methods. Whether those methods are simple or complicated, their output has to be made understandable to whoever needs it, be it the patient choosing a doctor, the healthcare professional improving their own performance, the manager overseeing their unit, or the regulator assessing the whole system. This book covers measuring quality, types of data, risk adjustment, defining good and bad performance, statistical monitoring, presenting the results to different audiences, and evaluating the monitoring system itself. It thereby brings all the issues and perspectives together in a largely non-technical format for clinicians, managers and methodologists.

Both authors moved from environmental epidemiology to health services research in 2002, following our involvement with the public inquiries into two big UK scandals: (1) paediatric cardiac surgery at the Bristol Royal Infirmary and (2) general practice by Harold Shipman, the country's most prolific serial killer and family doctor. Applying what we had learned from our work to those public inquiries, Alex Bottle devised the details of our national hospital monitoring system that helped uncover a third scandal: the high death rate at Mid Staffordshire NHS Hospital Trust. Our unit's research focuses on developing indicators of quality and safety and care and understanding performance variations in primary care and between surgeons and hospitals in a range of specialties. We are also involved in quality improvement projects and work with psychologists to understand how clinical staff and managers use (or ignore) performance information. Alex conceived this book as a way to bring together all the elements of performance monitoring, with examples from around the world. We were also keen for the algebra content to be as close to zero as possible, which it never is in statistics guides, not because writing formulae is fiddly, but because many statistical issues need to be understood by everyone interested in performance.

Alex Bottle
Paul Aylin
Imperial College London
United Kingdom

Authors

Alex Bottle is a senior lecturer in medical statistics in the School of Public Health at Imperial College London and methodological lead at the Dr. Foster Unit there. His background is in statistics and epidemiology, and he researches the use of large databases to measure and monitor the quality and safety of healthcare, with particular interests in risk adjustment, statistical process control and modelling multiple health service contacts in patients with chronic diseases. His fruitful collaborations with a wide range of clinicians, from anaesthetists to urologists, as well as with quality improvement scientists and the media, led him to conceive and write this book to bring all these worlds together.

Paul Aylin is a professor of epidemiology and public health at the School of Public Health at Imperial College London and leads the Dr. Foster Unit. After his medical training, he specialised in public health medicine and then spent three years at the Office for National Statistics in London before joining Imperial College in 1997. He became a Fellow of the Faculty of Public Health Medicine in 2001. He is a non-executive director at West London Mental Health NHS Trust. His research explores variations in indicators of quality and safety using routinely collected administrative data, including out-of-hours hospital mortality, the August (or July) effect, cancer care and indicators for late diagnosis. He has been an expert witness for both the Bristol Royal Infirmary and the Shipman public inquiries in the United Kingdom.

1

Introduction

> Between the health care we have and the care we could have lies not just a gap, but a chasm
>
> – Institute of Medicine
>
> In God we trust; all others must bring data
>
> – W. Edwards Deming

1.1 The Need for Performance Monitoring

Healthcare practitioners are not all created equal. Healthcare organisations are not all equally effective. This matters to everyone. As patients, we want to know whether the people treating us are giving us the best chance of better health. As healthcare professionals, we want to know that the care we offer is of a high standard. The government wants to know whether it's getting good value for its often considerable healthcare budget.

Performance monitoring has been driven by both scandal and science. Some high-profile failures such as Mid Staffordshire NHS Hospitals Trust in England and King Edward Memorial Hospital in Australia have piqued the public's interest. Since the 1990s, the United Kingdom alone has had three public inquiries over what went wrong and why it was not detected earlier. Landmark studies and reports of the considerable apparent dangers of being in hospital helped transform the famous instruction to first do no harm ("Practice two things in your dealings with disease: either help or do not harm the patient") into a global patient safety movement (Institute of Medicine, 2000). It's been known for some time that healthcare regularly falls short of its own standards in terms of not following clinical guidelines as often as they should be. In the words of the Institute of Medicine, the difference "between healthcare we have and the care we could have is not just a gap but a chasm" (Corrigan et al. 2001). This has been found in all sectors and all countries (Lee and Mongan 2012). As a result, healthcare has looked to other fields such as aviation and manufacturing to reduce the risk of harm and to raise standards in general. As the range, complexity and healing power of modern medicine have expanded, patients' expectations have risen. The financial costs of healthcare systems around the world escalate, and each system tries to do more with less. The system also needs to justify costs in terms of what it "produces" and how well it does it. All of this means that monitoring performance is needed more than ever.

1.2 Measuring and Monitoring Quality

Judging how well the healthcare system performs – measuring and monitoring the quality of the system and its many components – is a crucial but difficult task. This book aims to help you do just that. We cover the two vital elements to this:

1. *Measuring quality*: What data sources to use, how to construct quality indicators and how to analyse the data to identify areas where improvement is needed – or where excellence has been achieved and can serve as a beacon for others.
2. *Presenting the results to different stakeholders*: How should we inform healthcare providers of their performance, and how should we show this to the public?

Regulation, accreditation, clinical audits and other quality improvement initiatives are common and on the increase internationally. They all require the use of a range of data sources and statistical methods, which we describe. Databases, particularly administrative ones, are growing in accuracy and detail across the world, allowing more sophisticated statistical analysis. This brings its own challenges, such as how to summarise large amounts of information in a digestible but still useful way. The fanciest, most technically ingenious statistical modelling will serve for nothing if its output can't be understood by those who need to act on it. Many countries now report healthcare performance publicly and have the challenge of making such information accessible to the public without losing (too much) rigour or important detail. In Chapter 11, we give some examples of how this is currently done in the United Kingdom and the United States. We then review the evidence on which presentation formats have been found to help or hinder understanding: which can prompt staff to make changes in their own practice or at their organisation, and which best help patients make sense of the complex information.

1.3 The Need for This Book

Assessing and monitoring performance – whether our own or that of our colleagues, our clinic, our department, our hospital or our country – needs to be done well. We want to know that most providers that are labelled poor performers actually do have problems because false alarms undermine confidence, waste resources on unnecessary audits and investigations and can cause unwarranted reputational damage.

Monitoring is complicated and has a number of components, each of which continues to be the focus of research. Current books and reports cover part of the picture, such as measurement or general analysis methods, and some have high algebra content. There is a need to bring together all these issues in one place in a relatively non-technical but comprehensive way. In addition to a full list of references that we cite, we also direct you at times to books, reports or websites where specific methods are covered in more detail.

The belief that most things can be improved is the basic idea behind continuous quality improvement. This is a management philosophy that organisations in many sectors use to reduce waste and increase efficiency and satisfaction among employees and customers. The Sino-Japanese word kaizen (improvement) was made famous in the West by the Toyota Production System, the forerunner of the more generic lean manufacturing (or just "lean"). As in other spheres, continuous quality improvement has been widely adopted in healthcare. It's a different analytical paradigm from that of experimental science and the randomised controlled trial, and therefore different statistical approaches are needed. Our emphasis is consequently on observational studies, which have been the focus of much recent methodological development.

1.4 Who Is This Book For and How Should It Be Used?

The topics that we cover include defining quality, choosing the data sources, constructing and presenting indicators, flagging unusual (mainly poor or deteriorating) performance and evaluating the monitoring system itself. It's not primarily about how to implement quality improvement projects in practice. Some of these topics are more statistical than others, but all need to be understood by methodologists – and also by all those with responsibilities for monitoring standards within their organisation or their country. Increasingly, comparisons are made across national borders, the focus of Chapter 10; some clinical registries, for instance, have participants from a range of countries as units aim to compare their performance against their peers overseas. The book is therefore aimed not just at statisticians but at all those who need to know how to measure and compare performance. This includes researchers, government agencies, regulators, health service managers, health service commissioners and quality improvement scientists.

Some sections are more statistical than others, but there are few formulae. We assume that the reader has the level of knowledge of statistics as obtained from an undergraduate nursing or medical degree course. The more technical sections are marked with an *asterisk in brackets*. The book will not cover "how to analyse data from first principles". It may be read from beginning to end in order or as a resource to dip into for guidance on, or an explanation of, specific issues. No chapter requires the prereading of another. A number of issues are relevant to more than one chapter, and there are some recurring themes. We aim to minimise repetition by referring the reader to another chapter. Some issues, however, such as choosing the unit of analysis and reporting, need to be discussed from different angles depending on the purpose of the chapter.

Within a healthcare system, there are usually several levels, from the individual practitioner up to a regional network. Given the will and the right data, monitoring may be done at any or all of these levels. Most work in performance measurement and monitoring has been done in hospitals, and much of the pioneering thinking was done in surgery such as by the great Ernest Codman, who collected detailed and systematic information on errors and outcomes in the early twentieth century. In the growing field of patient safety in particular, primary care still lags well behind secondary care. For this reason, many of the examples in the book are taken from surgery and from other activity in hospital. Another recurring theme is death, which is of particular interest to everyone and naturally has been the subject of much research. Statistically, this means that there is more emphasis on handling binary outcome measures than continuous ones. However, the issues that we cover also relate to outcomes other than death, to healthcare workers other than surgeons, and to settings other than the hospital. We will use the term *unit* to refer to anyone or anything who deals with patients and whose performance is already or could be monitored. Among others, this could be the family physician, health visitor, pharmacist, optometrist, dentist, nurse, surgeon, midwife, physiotherapist, anaesthetist, clinical psychologist, local clinic, general practice, community hospital, care home, acute hospital, tertiary centre or network – or even the whole country.

After this section a list of abbreviations follows. Although each of them will be expanded in the relevant sections when they are used, we've provided some common ones here, especially if they appear in several places. Each chapter begins with its aims. References for each chapter are given just before Appendix A, which describes in non-technical terms some of the main statistical terms used such as regression, the p value and false alarm rate.

Common Abbreviations Used in the Book

Organisations and Their Main Country of Origin

ACHS   Australian Council on Healthcare Standards (Australia)
ACSQHC Australian Commission on Safety and Quality in Health Care (Australia)

AHRQ	Agency for Healthcare Research and Quality (USA)
CMS	Centers for Medicare & Medicaid Services (USA)
CQC	Care Quality Commission (UK)
HCFA	Health Care Financing Administration (USA)
HES	Hospital Episode Statistics (UK)
HPA	Health Protection Agency (UK)
IHI	Institute for Healthcare Improvement (USA)
IoM	Institute of Medicine (USA)
JAMA	Journal of the American Medical Association (USA)
NCQA	National Committee for Quality Assurance (USA)
NHS	National Health Service (UK)
NICE	National Institute for Health and Care Excellence (UK)
NPSA	National Patient Safety Agency (UK)
NQF	National Quality Forum (USA)
NSQIP	National Surgical Quality Improvement Program (USA)
OECD	Organisation for Economic Cooperation and Development (mostly Europe)
RCP	Royal College of Physicians (UK)
STS NCD	Society of Thoracic Surgeons National Cardiac Database (USA)
VA	Department of Veterans Affairs (USA)
WHO	World Health Organisation

Data and Statistical

ANN	Artificial neural network
CAHPS	Consumer Assessment of Healthcare Providers and Systems
CBA	Cost–benefit analysis
CEA	Cost-effectiveness analysis
CUA	Cost–utility analysis
CUSUM	Cumulative sum
DRG	Diagnosis-Related Group
EB	Empirical Bayes (estimate)
EHR	Electronic health record
FDR	False discovery rate
GEE	Generalised estimating equation
HCAI	Healthcare-associated infections
HEDIS	Health Plan Employer Data and Information Set
HES	Hospital Episode Statistics
HRG	Healthcare Resource Group
HSMR	Hospital standardised mortality rate
ICC	Intraclass correlation coefficient
ICD	International Statistical Classification of Diseases and Related Health Problems
KPI	Key performance indicator
MAR	Missing at random

MCAR	Missing completely at random
MOR	Median odds ratio
NRLS	National Reporting and Learning System
OR	Odds ratio
POA	Present on admission
PPV	Positive predictive value
PREM	Patient-reported experience measure
PROM, PRO	Patient-reported outcome measures or outcomes
PSI	Patient safety indicator
PSM	Propensity score matching
QOF	Quality and outcomes framework
RCT	Randomised controlled trial
RD	Regression discontinuity
RR	Relative risk
SCV	Systematic component of variation
SEM	Structural equation modelling
SES	Socioeconomic status (called deprivation in the United Kingdom)
SMR	Standardised mortality ratio
SPC	Statistical process control

Other

AIC	Akaike information criterion
AMI	Acute myocardial infarction
BP	Blood pressure
CABG	Coronary artery bypass graft
COPD	Chronic obstructive pulmonary disease
ECHO	European Collaboration for Health Optimization
GP	General practitioner or general practice (UK)
ITU, ICU	Intensive therapy/care unit
LMIC	Low- and middle-income country
MSE	Mean squared error
P4P	Pay for performance
ROC	Receiver operating characteristic
VTE	Venous thromboembolism

Acknowledgment

Alex Bottle thanks his sister, Janet Pitcher, for the time and effort she put into improving this chapter.

2
Origins and Examples of Monitoring Systems

Aims of This Chapter
- To outline high-profile failures of medical providers that have influenced how performance monitoring is done
- To provide some examples of performance monitoring systems
- To outline the different viewpoints of those interested in healthcare performance

2.1 Origins

Performance monitoring of healthcare is not a new idea. Back in Babylonia in around 1700 BC, King Hammurabi adapted existing cuneiform edicts into his Codex that had several important concepts we would recognise today: physician fees for treatments such as general surgery, eye surgery and setting fractures; objective outcome measurement standards to ensure quality of care; and, as the other side of pay-for-performance, punishments for failure. As an account of the Codex in Managed Care puts it, his edicts left no margin for error: healthcare providers had to be flawless or lucky (Spiegel 1997). §218 of the Codex states that

> If a physician operate on a man for a severe wound with a bronze lancet and cause the man's death, or open an abscess in the eye of a man with a bronze lancet and destroy the man's eye, they shall cut off his fingers.

Hammurabi's system also included a data collection mechanism for the patient's ailments, treatments and outcomes. We've moved on from his clay tablets, with the great nineteenth-century nursing reformer and epidemiologist Florence Nightingale who argued for a standardised collection of hospital mortality data. Such data would ensure that subscribers' money was being put to good and not, in her words, "mischief". Although best known for her nursing endeavours during and following the Crimean War, in 1858, Nightingale became the first female member of the Royal Statistical Society in the same year as publishing her seminal "Notes on matters affecting the health, efficiency and hospital administration of the British Army" (Nightingale 1858). In 1863, she published her "Note on Hospitals", where she stated that uniform hospital statistics might "... enable us to ascertain the relative mortality of different hospitals as well as of different diseases and injuries at the same and at different ages, the relative frequency of different diseases and injuries among the classes which enter hospitals in different countries, and in different districts of the same country". In fact, annual hospital mortality statistics were published in the *Journal of the Statistical Society of London* as early as 1862 for some London and provincial institutions.

An alternative approach, proposed by Ernest Amory Codman a few decades later, involved the detailed collection of outcomes of surgical cases and a coding system review to record why "perfection" had not been obtained. Codman was a Boston surgeon who in 1910 started his campaign for this system to systematically follow up surgical patients as the central concept behind his "End Results System". In 1914, he resigned his post at Massachusetts General Hospital after starting his own hospital for which he published his surgical outcomes. From 1911 to 1916, there were 337 patients discharged from Codman's hospital, for whom he recorded 123 errors. These were all published in the hospital's annual report – and he challenged the major hospitals of the country to do the same. His autobiography also disclosed his personal income. Unfortunately, his own hospital closed in 1918, but he continued to work on the standardisation and accreditation of hospitals, with his legacy contributing to the creation of The Joint Commission in the United States. For more details of his life and work and for further reading on this great man, see the piece by Neuhauser (2002).

More recent interest in performance measures has been driven by at least two factors: healthcare scandals and a need for greater efficiency because of rising healthcare costs. Perhaps a third principal factor is the wider trend towards medical transparency and openness in the medical profession. But first, scandals.

2.2 Healthcare Scandals

In the United Kingdom, it was problems at a children's heart surgery unit at the Bristol Royal Infirmary that drove performance monitoring up the political agenda. Since the mid-1980s, Bristol Royal Infirmary had been designated a special centre for child heart surgery. In 1988, an anaesthetist called

Steve Bolsin joined the cardiac team. He began to notice that cardiac procedures seemed to take longer than in other units, and he also suspected higher death rates. He started carrying out his own audit of death rates which confirmed his suspicions, and in 1992 he approached a Department of Health official about his concerns. No action was taken. In 1994, after completing his audit, he approached another Department of Health official, who, after reading it, called for a suspension of the particularly complex "switch" cardiac procedure. Despite being advised not to go ahead with these operations, the surgical team at Bristol operated on an 18-month-old baby who subsequently died on the operating table. This led to an external team visiting the unit who, after examining Dr. Bolsin's figures, concluded that Bristol was a "high-risk unit" and there had probably been "high-risk mortality". After complaints from local parents that Bristol doctors misled relatives about their success rates, the General Medical Council, which is the body responsible for standards in medical education and practice across the United Kingdom, carried out its own investigation. This led to the two surgeons involved and the medical director of the hospital being found guilty of serious professional misconduct. Furious at the fact that only two of the three doctors involved had been struck off the medical register, the then Health Secretary announced a public inquiry, led by Professor Ian Kennedy. After two years, the inquiry published its report. Key among its findings was that after a commissioned analysis of the mortality rates (Aylin et al. 2001), it appeared that the Bristol unit did indeed have a significantly high mortality rate in children aged under one year. There seemed to be no shortage of data supporting this, and indeed one of the conclusions of the inquiry was that Bristol "was awash with data from one source or another". The report called for greater transparency and that information on how hospital trusts, doctors and services within them perform compared with others should be made available to the public.

Not long after the Bristol Inquiry, the United Kingdom had another scandal. The need for monitoring healthcare performance became more urgent. This one concerned Harold Shipman, a general practitioner (GP) working in Hyde, Manchester. After a relative of one of his victims raised the alarm as a result of a clumsily forged will that named Shipman as the main benefactor, Shipman was arrested and subsequently convicted of murdering 13 of his patients through lethal injections of diamorphine. This led to a public inquiry which, through a review of casenotes of deaths occurring in patients of Shipman, later came to the conclusion that he had murdered at least 235 of his patients. At the peak of his murderous medical career, Shipman was killing almost a patient a week. One of the obvious questions that the inquiry asked was how a practising doctor had got away with murdering so many patients over such a long period without anyone noticing. The inquiry commissioned us (this book's authors) to examine statistical techniques for detecting unusual performance using data on deaths in patients registered with over 1000 GPs (one of whom was Shipman, though we didn't know which data were his until after the analysis). We used statistical process control charts (see Chapter 9) and did spot Shipman, along with a few other GPs and practices, with high patient mortality rates (Aylin et al. 2003).

Sadly, the United Kingdom is not alone in this. In 1999 in Australia, the chief executive of the King Edward Memorial Hospital (KEMH) in Western Australia expressed concerns over obstetric and gynaecological services at his hospital to the Metropolitan Health Service Board. As a result of these concerns, a formal expert review was commissioned to look at the "quality of clinical care provided by the obstetric and gynaecological service". Two independent experts from outside Western Australia reviewed services at the hospital and were critical of many significant aspects of care provided at KEMH. On seeing the report, the Honourable John Day, Minister for Health at the time, established an inquiry into services at the hospital chaired by Neil Douglas, which expanded on the expert review. One of the inquiry's more serious findings was the inadequate supervision of junior medical staff, particularly by the hospital's consultants. As an example, it found that a junior doctor or midwife was the most senior person who provided care at the crucial times in 70% of high-risk obstetric cases. Because of the lack of comparative data available to the inquiry, it established a consortium to provide information on outcomes using routinely collected perinatal, hospital morbidity and neonatal data. While acknowledging that KEMH had a higher proportion of more complex cases than other hospitals and that some of the sources of data used were primarily for administrative purposes, the consortium concluded that their findings were sufficiently robust to identify major differences among hospitals. They recommended that the hospital further investigate its high rates of stillbirths and obstetric interventions. It found a relatively large number of hysterectomies following post-partum haemorrhage and a high proportion of women transferred to the Adult Special Care Unit during admissions for laparoscopic procedures. The consortium also recommended improving the quality and completeness of data collected at the hospital (particularly morbidity data) and that KEMH should maintain obstetric, perinatal and gynaecological services outcome data. The Douglas Inquiry was considered a wake-up call to the Australian government, chief executives, managers and clinicians. It defined a clear imperative for rigorous data comparison, as well as effective incident monitoring and mortality review systems. The Australian Council for Safety and Quality in Health Care commissioned a report to bring out wider lessons from the Inquiry (ACSQHC 2002). The report noted a number of similarities with the findings from the Bristol Inquiry and that, despite there being many examples of exemplary care and staff trying to overcome clinical and management problems, the hospital lacked effective systems to report and respond to performance problems.

In Queensland, the conduct of a general surgeon, Dr. Jayant Patel, working at Bundaberg Base Hospital (BBH), was brought to light in 2005 after being named in the Australian Parliament by a member who had in turn been briefed by a whistleblower. Patel had been appointed as the senior medical officer at BBH in April 2003 and later became the director of surgery. However, he failed to disclose to his appointment panel that in 2001 he had been forced to surrender his New York State licence to practice and that in his most recent post in Oregon his clinical activity had been restricted by the Oregon Board of Medical Examiners over concerns about his surgery. During his two years at BBH, he alarmed both doctors and nurses with a catalogue of errors and poor practice. After he was named in parliament, Patel fled the country. An interim report produced by a Commission Inquiry in 2005, whose terms of reference were admittedly limited, found that Patel had lied on his application and that, given that he was not qualified to operate on some of the patients that had subsequently died, he should be investigated for murder. In 2006, a warrant was issued for Patel's arrest on 13 charges including manslaughter, grievous bodily harm and fraud, and after extradition from the United States he was found guilty of manslaughter and sentenced to seven years for each charge. However, his convictions were quashed on appeal and a retrial was ordered. Eventually, Patel was convicted of fraud only in 2013 and received a two year suspended sentence.

In the United States, the driver for closer scrutiny of healthcare performance was not so much a kneejerk reaction to specific scandals but arose more from a general observation of the variability of outcomes between hospitals and geographical areas. This was seen first in an early type of hospital profiling released by the Health Care Financing Administration in 1986. The first annual report provided hospital-level data for 17 diagnostic and procedure groups. However, it was heavily criticised for not taking adequate account of differences in casemix (patient risk – see Chapter 6), and the reports were eventually suspended in 1994. In an early precursor of the Dartmouth Atlas Project, a long-running project that uses Medicare data to show variations in resource use across the United States, Wennberg et al. (1987) compared per capita expenditures for inpatient care in the demographically similar populations of Boston and New Haven and found a twofold variation. Meanwhile at a specialty level, the Society of Thoracic Surgeons responded by establishing their National Cardiac Database (STS NCD) in 1989 as an initiative for quality improvement and patient safety, with voluntary online outcomes reporting by participants since 2011. Other analyses of risk-adjusted cardiac surgical outcomes began to show more variability than expected due to chance alone. In 1989, the New York State Department of Health began collecting detailed clinical data on all coronary artery bypass surgery (CABG) cases performed in that state, and risk-adjusted hospital- and surgeon-level rates were derived. In 1990, a New York hospital was suspended from performing CABGs, partly because their observed mortality significantly exceeded the expected rate. Following a 1991 lawsuit, New York officials were required to publicly release surgeon-specific outcome data under the Freedom of Information Act, and the state has voluntarily released annual figures since 1992. The effectiveness of this monitoring has been questioned. Supporting the monitoring, an analysis of rates between 1989 and 1992 found a reduction of 41% in risk-adjusted mortality (Hannan et al. 1994), concluding that much of this could be due to the monitoring and reporting system. In contrast, others proposed that changes in patient selection were the main reason for this apparent improvement, either by transferring high-risk patients to other states or not operating on them at all (Green and Wintfeld 1995). However, a later analysis that benefitted from using national trends as a comparison for New York State found that New York's post-CABG mortality fell by more than that of the country as a whole, with no evidence of change in out-of-state transfers, despite an increasing risk profile (Peterson et al. 1998). This debate over the reasons behind the mortality fall illustrates the importance of employing an appropriate study design that considers the potential for any intervention to have unintended consequences (Chapters 4 and 12).

2.2.1 Responses to the Scandals

The fires of scandal in the United Kingdom and Australia and self-analysis by clinicians in the United States highlighted the need for systems to monitor healthcare performance. Different countries responded in different ways. In the United Kingdom, the Bristol Inquiry was widely reported as being a pivotal moment when many of the most fundamental changes to how the National Health Service and organisations and individuals (including doctors) within it are regulated and monitored. In 2001, the Department of Health published the widely derided star ratings for hospitals. This system gave hospitals, primary care services and ambulance services a score of between zero and three stars derived from about 50 kinds of targets. Its critics argued that the ratings gave no indication of the quality of care patients could expect in their local hospital. Instead, the ratings relied on self-reported adherence to process-oriented targets such as cleanliness of wards and the time taken to see patients. The system of targets was used for performance management of public services and has been likened to the Soviet regime "which was initially successful but then collapsed", although it has been acknowledged that services improved following the introduction of these targets. The Healthcare Commission, then responsible for assessing the quality of National Health Service (NHS) healthcare, replaced these ratings in 2006 with an "annual health check". These incorporated data beyond simple targets and sought to make better use of available data. Recommendations from the Shipman Inquiry resulted in a revalidation program for all senior doctors, including GPs, which requires them to be assessed every five years on their fitness to practice. In addition to public systems for performance monitoring, commercial companies such as Dr. Foster Intelligence started to publish analyses of hospital mortality, the HSMR, developed by Jarman et al. (1999). Their first Hospital Guide was published in 2001 and for the first time allowed members of the British public to see hospital mortality rates. Several other countries later followed suit.

Application of the statistical process control methods seen in the Shipman analysis to hospital data led our team to develop a national hospital mortality surveillance system in England. This uses hospital administrative data, updated monthly, to alert hospitals of unusually high rates of mortality for individual diagnosis and surgical procedure groups (Bottle and Aylin 2008a). The value of this system was demonstrated just months after it was set up in 2007. It generated a number of statistical alerts triggered by high death rates at Mid Staffordshire NHS Foundation Trust, which prompted an investigation by the Healthcare Commission. Their investigation found appalling standards of care, particularly for patients admitted as an emergency. The findings of this initial investigation led to a further independent inquiry and a subsequent more comprehensive public inquiry – the country's third in 10 years. The latter confirmed the Healthcare Commission's initial findings, and its recommendations have had far-reaching consequences for the use of performance data and the monitoring of healthcare organisations and professionals in the United Kingdom. The Care Quality Commission, the successor to the Healthcare Commission, has developed an inspection regime which uses surveillance data from a variety of sources including avoidable infections (such as *Escherichia coli* infections), notifications of deaths, severe and moderate harm and abuse, reporting of "never events" (see Section 4.3) and mortality rates in various diagnosis and procedure groups. The identification of Mid Staffordshire was different from the cases of Shipman, Bristol in the United Kingdom and King Edward Memorial Hospital and Bundaberg Base Hospital in Australia in that Mid Staffordshire was identified partly through a surveillance system specifically set up to monitor performance of healthcare.

The Australian response to their scandals was to review their accreditation process, making systems more accountable to government and the public. The Australian Commission on Safety and Quality in Health Care (ACSQHC) was set up in 2006 and took responsibility for the determination of service standards, latterly with the introduction of mandatory National Safety and Quality Health Service Standards in 2013 for all hospitals and day procedure services. The Australian Health Service Safety and Quality Accreditation Scheme coordinates a number of approved accrediting agencies to rate services against these standards. The National Health Performance Authority was set up as an independent agency in 2012 to report on the performance of hospitals and primary healthcare organisations across Australia and has begun to report on the limited number of indicators such as *Staphylococcus aureus* bacteraemia and hand hygiene. State-level organisations such as the New South Wales Bureau of Health Information publish performance profiles. These include detailed information on indicators such as time to treatment for patients presenting to the emergency room and 30-day mortality for a number of specific conditions.

2.3 Examples of Monitoring Schemes

In the United States, a number of performance monitoring programs arose from the pioneering state-based cardiac reporting systems or surgical society initiatives mentioned earlier. In 1995, the National Committee for Quality Assurance (NCQA) released the first-ever report card on health plan performance in the United States. The NCQA now compiles the Health Plan Employer Data and Information Set (HEDIS). This tool is used by over 90% of health plans and by a number of organisations to measure performance at both the plan and provider levels. It consists of a number of measures across five different domains of care, which are updated each year. The majority of indicators are focused on outpatient care, and it's fair to say that the most measures (even the so-called effectiveness measures) are measures of process, with scant information on actual patient outcomes (we discuss types of measure in Chapter 4). The Centres for Medicare & Medicaid Services has collaborated with the NCQA to launch the Medicare Hospital Outcomes Survey (HOS) program, which attempts to summarise physical and mental health outcomes. Some positive association has been demonstrated between HEDIS and HOS in diabetes care (Harman et al. 2010), but another study comparing HEDIS and patient-orientated asthma outcomes concluded that that there was no association (Lim 2008).

The Joint Commission (formerly The Joint Commission on Accreditation of Healthcare Organizations) was founded in 1951. As an independent not-for-profit organisation, it accredits more than 20,000 healthcare organisations in the United States. One of the main reasons for such an inclusive membership is that the Centres for Medicare & Medicaid Services (CMS) requires all hospitals and a range of other healthcare providers to be accredited by The Joint Commission in order to participate in the Medicaid and Medicare health insurance schemes. In 1997, in an attempt to standardise performance measurement, The Joint Commission introduced outcomes and other indicators into the accreditation process, dubbed the ORYX initiative. In 2010, it categorised its performance measures into accountability and non-accountability measures, with more emphasis on the accountability measures. Hospitals can choose from measure sets, but additionally the Commission mandates data collection for core measurement sets for all general medical/surgical hospitals. The core measure sets include indicators on acute myocardial infarction (AMI), heart failure, pneumonia and a surgical care improvement project. Anyone can access these performance results online at http://www.qualitycheck.org.

The Leapfrog Group was launched in 2000 and is an employer-based coalition, publishing a set of measures focusing on hospital quality and safety practices. Hospitals submit information voluntarily using the Leapfrog Hospital Survey tool. Leapfrog also assigns a hospital safety score based on error prevention, accidents, injuries and infections. The results are freely available on their website at http://www.leapfroggroup.org and updated monthly.

Origins and Examples of Monitoring Systems 15

The 2015 survey had eight sections other than the basic hospital information: use of computerized physician order entry to prevent medication errors; evidence-based hospital referral for four high-risk surgical procedures; maternity care for normal and high-risk deliveries; ICU physician staffing, covering whether physicians are certified in critical care; safe practices score, covering adherence to 8 of the 34 National Quality Forum's safe practices; managing serious errors, relating to the prevention of hospital-acquired infections and other conditions and the hospital's Never Events policy (see Section 4.3 for measuring patient safety); bar code medication administration, new in 2015, for preventing medication errors; and resource use for common acute conditions.

The Agency for Healthcare Research and Quality (AHRQ) is a federal agency within the Department of Health and Human Services in the United States. The AHRQ has developed a range of quality indicators based on inpatient administrative data sets that are used within a growing number of report cards. The indicators cover four areas: inpatient, prevention, patient safety and paediatric care. The AHRQ claims that the indicators can be used to "highlight potential quality concerns, identify areas that need further study and investigation, and track changes over time". The Consumer Assessment of Healthcare Providers and Systems (CAHPS) program has also been developed by the AHRQ and is an initiative to support the assessment of consumers' experiences with healthcare. The survey-based program asks consumers and patients to rate their experiences within healthcare within domains such as the communication skills of providers and ease of access to healthcare services. See Section 5.6 for more examples of surveys.

Given the expanding international interest in performance monitoring, what are the goals of doing so? What do we hope to gain for our efforts and investment?

2.4 Goals of Monitoring

2.4.1 Accountability

Of the long list of issues addressed by the Bristol Inquiry, one of the key issues was the need for greater accountability (Kennedy 2001). Patients have also argued that public healthcare systems should be seen to be accountable to the taxpayer. Accountability in healthcare has been defined by the think tank the King's Fund as the requirement for organisations to report and explain their performance (Maybin et al. 2011). They suggest a number of different types of accountability, with perhaps differing reliance on healthcare performance indicators. The first type is scrutiny, for example, of hospitals by local overview and scrutiny committees. The second is management, both within and between organisations. The third is regulation. As an example, the United Kingdom's Care Quality Commission regulates health and social care services and, combined with inspections, uses a variety of sources of information and indicators in their "Intelligent Monitoring" system. Another type of accountability is required for contracting purposes, such as for commissioning or purchasing services.

2.4.2 Regulation and Accreditation

Regulation occurs in all developed countries in various ways and has been defined in healthcare as "any set of influences or rules exterior to the practice or administration of medical care that imposes rules of behaviour" (Brennan and Berwick 1996). A review by the Health Foundation (Sutherland and Leatherman 2006) of whether regulation improves healthcare noted that regulation has three key and potentially conflicting purposes:

1. To improve performance and quality
2. To provide assurance that minimally acceptable standards are achieved
3. To provide accountability both for levels of performance and value for money

The Health Foundation report classified approaches to regulate institutions into directive approaches (target setting and standard setting) and external oversight (accreditation and inspection). The two are often used in tandem. The report covers the regulation of professionals, which is done largely by licensure of individual practitioners, and of healthcare markets. In Chapter 9, we will cover targets and standards in the context of trying to identify "unusual" performance, so here we'll consider accreditation: first what it is and second whether it works.

Accreditation goes back about 100 years when the American College of Surgeons established a set of minimum standards for training posts in surgery. It's used mainly in hospitals but also in non-hospital settings. Program participation is generally voluntary and involves a formal assessment of whether the institution has met predetermined, published standards that are set to be achievable (and are sometimes criticised as not very demanding) and to encourage continuous quality improvement. Non-participation can lead to a competitive disadvantage, and the voluntary nature is blurred in countries that use accreditation as a component in licensing, for example. Increasingly, governments have tended to wrest control of schemes through health reforms and mandatory requirements. Accreditation can be considered to be an external review of quality with four main components:

1. Based on written standards
2. Reviews conducted by professional peers
3. Administered by an independent body
4. Aims to encourage organisational development

In the United States, The Joint Commission evaluates and accredits more than 20,000 healthcare organisations and programs. As mentioned earlier, a major incentive for accreditation in the United States is that it is one of the requirements for participation in the Medicare scheme. In the United Kingdom, there is no national equivalent to The Joint Commission as performance for healthcare providers is regulated through the Care Quality Commission. However, some private providers of healthcare do choose to join voluntary accreditation schemes.

So, does accreditation improve outcomes?

Most of the evidence on what works is observational and from the United States. The 2006 Health Foundation report concluded that the use of inspection, targets and standards has led to improvements in quality (unintended consequences notwithstanding). However, "[while] within systems that rely heavily on accreditation, accredited organisations generally provide higher quality care..., there is no evidence to suggest that accreditation has secured improved quality". The financial costs for the institution to participate are considerable, and it has been argued that accreditation is not a good use of money in low- and middle-income countries (LMICs). Ovretveit (2002) argues that "if the money used for the accreditation system was invested in quality methods to improve immunisation programs or drug supply logistics, many more lives would be saved and the changes would be sustainable".

A 2008 systematic review found 66 papers and categorised them into 10 topics: professions' attitudes to accreditation, change promotion, organisational impact, financial impact, quality measures, program assessment, consumer views or patient satisfaction, public disclosure, professional development and surveyor issues (Greenfield and Braithwaite 2008). There were few studies on the costs of accreditation, but they were high. The relationship between quality measures and accreditation was described as "complex", and the authors summarised the relations with quality measures as "inconsistent". In several studies, no relationship was found between a specified quality measure and accreditation. More positively, Collopy et al. (2000) reported that a significant majority of organisations responded to feedback in the Australian Council on Healthcare Standards (ACHS) accreditation program to improve both the processes and outcomes of patient care. In 1993, the ACHS introduced indicators into this program, unusual in its formal involvement of providers in the development process and in the scope of the clinical areas covered. In July 2002 in the United States, The Joint Commission implemented standardised performance measures that were designed to track the performance of accredited hospitals and encourage improvement in the quality of healthcare. A much-cited analysis of 18 of these mostly process measures showed consistent improvements following quarterly feedback in 15 of them (Williams et al. 2005). Relationships between consumer views or patient satisfaction and accreditation were largely unexplored, but no significant associations were found in the few studies that did examine this angle.

In contrast, a 2011 systematic review reached a quite different conclusion and came out more strongly in favour of accreditation (Alkhenizan and Shaw 2011). It was more recent than the 2008 one but strictly limited to health services accreditation rather than include other types of accreditation programs, such as those of medical education, as the earlier one had done. Of the 26 studies included, 10 evaluated the impact of a general accreditation program on the overall performance of hospitals, 9 studies evaluated the impact of a general accreditation program on a single aspect of hospital performance, and 7 studies evaluated the impact of subspecialty accreditation programs. Among the first type was the only published randomised controlled trial in this sphere (Salmon et al. 2003). In South Africa, 20 randomly selected public hospitals, stratified by size, were selected. Half were randomised to the accreditation program in 1998, the others serving as controls; the largest dropped out during the study, and so a control hospital was dropped to retain balance. Survey data from the Council for Health Services Accreditation of Southern Africa (COHSASA) program were used, along with data on eight indicators collected by an independent research team. After about two years since accreditation began, intervention hospitals significantly improved their average compliance with COHSASA accreditation standards, while no appreciable increase was observed in the control hospitals. Nonetheless, only one of the intervention hospitals achieved accreditation status at the end of the study period, while two others achieved pre-accreditation status. However, with the exception of nurse perceptions of clinical quality, the independent research team observed little or no effect of accreditation on seven other indicators of quality, including patient satisfaction. It has been suggested as an explanation for this that these hospitals had already made considerable progress that was not captured because the first round of the quality measure collection occurred some 10 months after the COHSASA baseline survey in the intervention hospitals, making it too late to be a true baseline. Other studies in the review support the authors' assertion that general accreditation programs improve the process of care provided by healthcare services and often also improve the outcomes. They note, however, the findings from a 2002 analysis of the National Committee on Quality Assurance (NCQA) and Health Plan Employer Data and Information Set (HEDIS) databases (Beaulieu and Epstein 2002). In that study, accredited plans had higher HEDIS scores but similar or lower performance on patient-reported measures of health plan quality and satisfaction. A substantial number of the plans in the bottom decile of quality performance were accredited, suggesting that accreditation does not guarantee high-quality care.

Finally in this section, we summarise a recent analysis from Denmark of accreditation by the first version of The Danish Healthcare Quality Programme for hospitals from 2010 to 2012 (Falstie-Jensen et al. 2015). Compliance was assessed by surveyors via an on-site survey and awarded the hospital as a whole either fully (n = 11) or partially accredited (n = 20) status. As relatively few studies had considered outcome measures, the researchers compared adjusted all-cause 30-day mortality and found 17% lower odds of death in fully compared with partially accredited hospitals (the difference was statistically significant). When reflecting on why this might have been, they make two important points. The first is that of course an association does not necessarily mean causation. In Denmark, the introduction of the accreditation program did not occur in a vacuum but in the presence of other quality improvement projects such as a Danish version of the Institute of Healthcare Improvement's 100,000 Lives Campaign (active from 2007 to 2009) and continuous indicator monitoring and auditing through clinical quality databases. In the same vein, it's possible that good compliance with accreditation standards could simply be a marker for high performance: hospitals that achieve good outcomes such as low mortality may be doing many things well, including meeting the accreditation standards, but they are not meeting those standards that enable them to achieve their good outcomes. The second point is the opposite: correlation might indeed mean causation. It's possible that the ability to effectively implement other quality improvement initiatives could in fact be a direct consequence of reorganising the hospital according to the standard required by the accreditation program. The authors were unable to tease apart these competing explanations with the data they had.

The financial costs of regulation are undoubtedly enormous. For example, Conover (2004) estimated these to be $339 billion in the United States in 2002, with only $170 billion recouped in savings attributable to it. There are also indirect costs such as those incurred by units preparing for accreditation visits. Such sums will always come under scrutiny, particularly when placed alongside the sparseness of measurable benefits in terms of money saved and patient outcomes, but we should remember that regulation also has the goals of accountability and transparency. These are valued by the public and reflected in their trust in the system and in healthcare providers. Of course, you can't easily translate such trust into any measure of patient outcomes or into financials. Overall judgments of whether regulation is "worth it" will inevitably stumble at this stage. In one sense, it's worth it if we as nations are willing to pay for it.

In summary, the pattern of findings is rather mixed and subject to more than one interpretation. As the Danish authors of the last study conclude, "More insights into the effect of accreditation on patient outcomes and processes of care, including the cost-effectiveness of this quality improvement strategy, are clearly needed". However, none of this has stopped their growing popularity, particularly in low- and middle-income countries.

2.4.3 Patient Choice

One of the drivers for publication of performance figures has been to help patients choose their healthcare provider. Reforms in the United Kingdom within the last 10 years, designed to increase efficiency and improve quality through competition, have been grounded on the principle of giving patients easy access to information about services and performance. Providing information to support choice is a major priority for the NHS and is enshrined in the NHS Constitution. The NHS Choices website was launched in 2007 and supports the idea of providing better information about competing services in order to enable patients to make informed choices about their services. Whether the availability of this type of information is actually useful for patients is still open to debate. A survey that asked patients which sources of information they consulted to help them choose reported that only 4% of patients who were offered a choice consulted NHS Choices. Most relied on their own experience or their GP's advice, with a strong preference for local services (Dixon et al. 2010). Similarly, work in the United States suggests a limited effect on patient choice of Consumer Assessment of Healthcare Providers and Systems performance information on health plans. This may be because of the complexity of the information on offer, which particularly penalises patients with adverse socioeconomic status. Because of this, patient choice could actually lead to inequalities.

A few studies have asked whether media reports of healthcare provider problems make a difference to patient behaviour. This is public reporting with a high profile and so might be expected to have more of an impact. In 2008, the Healthcare Commission in England published a report, "Towards Better Births", with patient and staff survey results (not ranked by performance). After the release of the patient survey results on a website called "Birth Choice UK" in November 2007, TV and newspapers picked up on the "10 best" and "10 worst" maternity care providers in the country. The BBC website included headlines such as "NHS maternity units falling short". Although data were released on the quality of maternity care for all hospitals, there was no media reporting of those outside the extreme 20, allowing a natural comparison group. A difference-in-difference analysis (see Chapter 12) found no statistically significant change in overall maternity admissions in the best 10 hospitals (+2.2%, $p = 0.40$ at 6 months) or the worst 10 hospitals (−2.8%, $p = 0.49$ at 6 months) during any period in the 36 months after public reporting compared with their matched comparison groups (Laverty et al. 2015). The public did not seem to have changed their behaviour.

The same research group looked at trends in elective (planned) admissions to three hospitals that were subject to high-profile investigations by the national regulator for England between 2006 and 2009 (Laverty et al. 2012). Such investigations often attract media attention. Two of these concerned the management of *Clostridium difficile*, an important pathogen in hospital-acquired diarrhoea; the other one concerned the high death rates at Mid Staffordshire that resulted in the public inquiry mentioned earlier. Analysis showed no impact on utilisation for two of the hospitals; in the third hospital – interestingly, this was not Mid Staffordshire – there were significant but short-lived falls in inpatient admissions, in day-case (outpatient) surgeries, and in numbers of patients coming for their first outpatient appointment, but the effects disappeared 6 months after the regulator's investigation was published.

2.4.4 Openness and Transparency

Despite the low use of information on healthcare performance in choosing services, independent monitoring of healthcare performance is considered necessary, and patients have argued that this information should still be in the public domain. Recommendations arising from a number of healthcare scandals have argued for greater openness and transparency. Professor Ian Kennedy in his Bristol Royal Infirmary Inquiry final report argued that "trust can only be sustained by openness", and that "openness means that information be given freely, honestly and regularly" (Kennedy 2001). He also argued that "informing patients must be regarded as a process and not a one-off event" and that "the public and patients should have access to relevant information". Sir Robert Francis reiterated the need for transparency in his report on Mid Staffordshire and argued that "true information about performance and outcomes [should] be shared with staff, patients and the public" (Francis 2013). We will return to the evidence around the public reporting of patient outcomes by surgeon and hospital in Chapter 11.

2.4.5 Quality Improvement

A common reason given for performance monitoring is that it drives improvement, either through patient choice or through competition. Both of these mechanisms involve the spread of information, often through some kind of public reporting to enable the public to vote with their feet if they don't like a given healthcare provider, or through feedback of report cards or other information tools. The Institute of Medicine's view is that public reporting of performance will increase transparency, accountability and quality (Corrigan et al. 2001). There is some evidence that this is the case. In a rare experimental design, Hibbard and Peters (2003) looked at groups of hospitals where a performance report (public Quality Counts report) for each hospital either was made publicly available, provided to hospitals in confidence, or was not provided at all. After a two year follow-up, and focusing on obstetric care, they found that after hospital size was adjusted for, the "public report" hospitals had significantly improved their scores over the "no report" hospitals. They also found that the hospitals that received private reports improved their performance scores over the "no report" hospitals. However, this widely cited study had significant limitations in that it restricted its analysis to two specialties, and only obstetrics provided any significant results. In a randomised trial of public reporting, Tu et al. (2009) found that publication of acute myocardial infection (AMI) and heart failure process measures of performance resulted in reductions in 30-day AMI mortality. However, systematic reviews of the evidence published before the study by Tu et al. had concluded that any association between public reporting and quality improvement is modest and generally limited to a few specialties such as cardiac surgery (Marshall et al. 2000; Fung et al. 2008; Shekelle et al. 2008).

A little more encouraging was the synthesis of 11 hospital-level studies in the review by Fung et al. (2008) that suggested that public reporting did at least stimulate quality improvement activity, even if the relation with improved effectiveness was inconsistent. As noted in Section 2.4.3, there's little evidence that patients are using this information as consumers to change the provider. A more recent Cochrane review of the evidence for improvement following publication of healthcare data included just four relevant studies and concluded that there was no consistent evidence for changes in consumer behaviour or improvements in healthcare following publication of performance data (Ketelaar et al. 2011). A narrative analysis of the evidence decided that "public reporting of comparative data does seem to have a limited role in improving quality, but the underlying mechanism is providers' concern about their reputation, rather than direct, market-based competition driven by service users" (Marshall and McLoughlin 2010).

There have been a number of reviews looking at the impact of performance feedback on individual clinicians' behaviour. Veloski et al. (2006) reviewed all empirical studies (including quasi-experimental designs) with data before and after feedback was given, limited to studies relating to physicians' clinical performance. Of the 41 studies that met their inclusion criteria, 70% reported positive effects of feedback on performance. Importantly, they found that feedback from an administrative unit or professional groups and a longer duration of follow-up period was associated with positive results. Of another 132 studies that examined the effect of feedback combined with other interventions such as educational programs, practice guidelines and reminders, 106 studies (77%) demonstrated a positive impact. A more recent review of 140 randomised controlled trials measuring effects on not just doctors but nurses and pharmacists found that audit and feedback generally lead to small but potentially important improvements in professional practice (Ivers et al. 2012). The review concluded that the effectiveness of audit and feedback "seems to depend on baseline performance and how the feedback is provided. Future studies of audit and feedback should directly compare different ways of providing feedback". We will see in more detail in Chapter 11 how the method of presenting performance information to staff and to the public affects how that information will be used. The relation between public reporting and quality improvement thus far has been a bit disappointing, but it's clear that the field is still learning how to maximise its potential.

2.4.6 Prevent Harm and Unsafe Care

The other side of the coin to driving improvement in care is preventing harm and unsafe care. Although the Hibbard paper found improvements of summary indices of adverse events following publication, this was within only two specialties. The evidence is generally limited. However, surveillance is one of the pillars of infectious disease control, and, with healthcare-associated infections (HCAIs), there is evidence of the use of performance monitoring associated with reduction in harm. Surveillance can be carried out through either mandatory reporting of infections or through point prevalence studies. Each has their advantages or disadvantages. Point prevalence studies allow collection of more comprehensive data and are useful in examining the epidemiology of infection in greater depth. The most recent Public Health England HCAI and Antimicrobial Point Prevalence Survey looked at data collected in English hospitals between September and November 2011 for a variety of HCAIs including respiratory, urinary tract, surgical site infections and clinical sepsis (Health Protection Agency 2012). Continuous surveillance allows the monitoring of trends and hence the assessment of the impact of interventions and control measures. Such systems can be mandatory (such as the system in Ontario for *C. difficile* and for methicillin-resistant *S. aureus* in England since 2004). It allows the long-term pattern of these infections to be analysed and can help identify hospitals with high rates of infections.

2.4.7 Professionalism

A Royal College of Physicians (2005) report on professionalism and medicine suggested that medical professionalism should include continuous improvement and excellence. One can argue that in order to ensure both of these, practising clinicians need to know how they are performing in relation both to their own past practice and compared with peers. Some measure of performance is therefore required for good professional practice. Recently, surgical outcomes have been published for some 5000 hospital consultants practising in the United Kingdom. There are plans for eventually publishing outcomes for all NHS consultants. The arguments made for this initiative have revolved around choice, transparency and quality improvement, but there has also been a strong moral argument that clinicians have a professional responsibility to be able to describe what they do and defend how well they do it. Sir Bruce Keogh, the medical director of the NHS, describes this as "the essence of professionalism."

2.4.8 Informed Consent

A strong argument has been made that, for the provision of truly informed consent to surgical patients, there needs to be public disclosure of comparative performance measures of surgeons. In order to provide informed consent, patients need all the available relevant information to hand. This would include the success (or failure) rate for a given procedure, which might include a number of measures of complication rates and patient outcome rates. Although the provision of some sort of estimate of risk to a patient is not in dispute, Clarke and Oakley (2007) go further, arguing that this should include information about the ability of a surgeon to perform the operation in the form of consultant-level performance measures. Others have argued against this, not on an ethical basis, but on the basis that it might make some surgeons risk-averse, reluctant to operate on complex cases in case they are seen to inflate their death rates. This was suggested early on as a potential problem for the New York CABG monitoring system mentioned earlier.

Performance monitoring has had its pioneers, such as Nightingale and Codman, and a range of motivations, such as professionalism and public scandals. Taking on myriad forms, there is some level of it in almost every healthcare system in the world, but many statistical, presentational and other aspects are still being developed and evaluated. We will now consider these in the remainder of the book.

3

Choosing the Unit of Analysis and Reporting

Aims of This Chapter

- To give examples of how observations can be "clustered" and why this matters
- To set out the issues and options when analysing and reporting clustered data
- To discuss the issue of attributing performance to a particular unit in a system

We need to distinguish between the unit of analysis and the unit of reporting as they have different but overlapping sets of issues. If we want to compare clinics in terms of their staffing levels or hospital wards in terms of their bed numbers, for instance, the unit of analysis and the unit of reporting are typically the same. Analysis, reporting and interpretation are relatively straightforward. For other measures, the units of analysis and reporting are always necessarily different. For example, many figures reported by hospital, such as risk-adjusted mortality rates for acute myocardial infarction (AMI), involve analysis at the patient level and then reporting by hospital. In these examples, there is no choice in the unit of analysis, but in others there can be. We first consider the issues regarding the analysis because if that isn't sound, then there's little point in the reporting. Some issues affect both. After describing the issues, we give some potential solutions. Lastly, we discuss the special problem of attributing performance to a given unit when multiple units have been involved in the patient's care. This is, of course, extremely common as patients are treated by multiple members of healthcare and other staff and can be transferred from one provider to another.

3.1 Issues Principally Concerning the Analysis

In specialties that cover multiple body sites such as rheumatology, ophthalmology and dentistry, should the analysis be based per person or per joint, eye or tooth (or even tooth surface)? In an assessment of the prescribing behaviour of physicians in general practice, should the analysis be based on prescribing at the level of the physician, the practice, the health authority – or at the level of the patient or even the prescription? As we now explain, the choice matters.

3.1.1 Clustering (*)

A common assumption made in many statistical methods is that the observations are independent of each other. In many situations, however, they will be correlated to some extent. A set of patients all managed by the same doctor will show more similarity in their response to treatment than the same set would if they were all managed by different doctors: their responses will be more highly correlated. This reduces the effective sample size of the analysis: it's as if we didn't have as many observations as we thought we had. Patients treated by the same doctor are said to be "clustered by" or "nested within" doctor. There may be more than one level of "clustering" such as patients by surgeon and surgeon by hospital or joints by finger and finger by patient. In principle, failure to account for this "intraclass correlation" can lead to the wrong conclusions. Such errors are common (Andersen 1990). In a monitoring system, it can lead to units wrongly labelled as "good" or "bad" when in fact their performance is average. In practice, the degree of correlation can be small and so will have a negligible impact on the results, but it's dangerous to assume this at the outset. In our experience, it is more often necessary to account for it in primary care than in secondary care. Box 3.1 provides some common examples of clustering.

Kerry and Bland (1998) consider the unit in cluster randomised controlled trials (RCTs) in primary care. Randomisation is done at practice level because there's a risk that intervention patients may "interfere" with control patients at the same practice. The unit of analysis should therefore also be the

---

BOX 3.1   COMMON EXAMPLES OF "CLUSTERING"

Patient – surgeon – hospital
Patient – insurer – hospital
Patient – primary care physician (GP) – clinical commissioning group
Patient – oncologist – hospital – cancer network
Joint – hand – patient – physician
Tooth surface – tooth – patient – dentist

practice, even though it incurs a loss in power compared with the case when interventions are done (and then analysed) at the level of the patient. This should also be done when the intervention is at the level of the primary care medical practitioner or practice nurse. In a similar vein, Bland and Altman (1994) discuss the scenario of multiple observations per patient and conclude that, because the patient is the sampling unit, the analysis should be at the patient level and take account of the correlation. They cite a study that used nearly 4000 observations on just 58 patients where those observations were analysed as if each came from a different person. By ignoring the correlation, a falsely small p value was obtained, and, therefore, a statistically significant difference was found where in fact none existed. The standard errors from a model that deals with the correlation can be several times larger than those from a model that doesn't.

There are several valid ways of dealing with this. The *simplest* is to make a summary measure of all the observations for each patient, perhaps by taking an average. This may be good enough, depending on how much important information is lost in the summary. If the measurements for the same person are highly correlated, then little will be lost. It may be important to note, however, whether the number of observations taken is due to how frequently the patient is monitored. Frequent monitoring can be a marker of disease severity, and this fact could be used in the analysis to adjust for patient risk. As usual, the details of analysis need to be thought through before the data collection if possible. In ophthalmology, investigators should consider whether to collect information on each eye. This will depend on how each eye is affected by the disease of interest and how they are treated. In a clinical trial, each eye may receive a different treatment, when the analysis is then of paired data (with eyes paired within a patient); if they have both received the same treatment, then this within-patient correlation needs to be accounted for (Bunce et al. 2014). With very high correlation between the two eyes, an average may be taken or one eye can be chosen at random to avoid bias. With non-normal, for example, binary outcomes, values corresponding to the "best" or the "worst" could be taken. Armstrong (2013) reviews the varied approaches in the literature and gives recommendations for this specialty, particularly within the clinical trial setting. For orthopaedics, some patients require that both hips or both knees be replaced, with each replacement having its own outcomes of interest. Of course, each replacement is subject to the same person-specific effects (more or less) and so are not independent.

Another way is to use regression followed by the derivation of "cluster-robust" standard errors. These do not make us specify a model for the within-cluster error correlation as some other methods do. However, they do assume that the number of clusters, rather than just the number of observations, is very large. For a national analysis of healthcare units, that condition is likely to be met. Otherwise, if there are, say, 50 or fewer clusters, then it won't. NB: there's no clear-cut definition of what constitutes "a large number". With few clusters, the method can introduce bias and wrong standard errors. The problem is worse if the clusters are not "balanced", that is, if they're not of the same size. There are some ways around this: for a technical discussion of the problem and some solutions, see Cameron and Miller (2015).

Alternatively, we can choose to explicitly account for the multiple observations. Analysis of variance (ANOVA) is a class of methods that partition the variation in the outcome variable of interest among its different sources, typically between and within groups. "Group" here would be the healthcare unit. At its most basic level, ANOVA generalises the t-test to compare means across more than two groups. It's most commonly used with experimental data where the sources of variation can be changed with each experimental run. It assumes a normally distributed continuous outcome variable and can't handle covariates. For performance monitoring, it's probably not of much use to us.

With just one degree of clustering, such as patients by physician or multiple measures for each patient over time, a common approach is to use so-called marginal or population-averaged models, which typically use generalised estimating equations to provide the output. These can be applied to both normal and non-normal outcomes. The resulting coefficients are often described as "population-averaged" because the model supposes that the relationship between the outcome Y and the covariate X is the same for all patients. For information on how this model deals with multiple observations per patient over time and missing values, see Carrière and Bouyer (2002). They also compare it with multilevel modelling (see later text). It's fair to say that both approaches have their pros and cons (and their proponents) and that there's no consensus on which is best in general (Mason 2001; Hubbard et al. 2010). The balance will be tipped towards one or the other depending on such factors as the goal of the modelling and who will need to interpret the output. Marginal models seem to be more useful and more used in epidemiology (Greenland 2000; Hubbard et al. 2010) than in health services research, where multilevel models seem to be winning the day. We'll return to this comparison in Section 7.5.

However, the most common and most powerful method for dealing with the problem is multilevel (aka hierarchical) modelling, which fully represents the variation in the data and so invokes more assumptions. For example, it allows the relation between the outcome and the covariate to differ between patients, unlike the population-averaged model just mentioned. This is discussed fully in Chapter 7, as different options are available both for specifying the model and for producing risk-adjusted outcomes by a healthcare unit.

Dependence between observations is not the only issue that can arise when considering the unit of analysis. The other key ones, summarised in Table 3.1 with their implications, are sample size, multiple testing, attribution and the modifiable areal unit problem (MAUP). We'll now go through each in turn.

TABLE 3.1

Main Methodological Issues When Considering the Unit of Analysis

Statistical or Methodological Issue	Explanation	Consequence of Ignoring It	References for Further Reading
Sample size and random variation	The larger the unit of analysis (the more aggregated it is), the more robust that unit's estimates and the greater the power to spot good and poor performers.	A monitoring system based on binary measures for small units (e.g., death rates for low-volume surgeons) may not distinguish between good and poor performers.	Physician offices or general practices (Landon and Normand 2008); Surgeon volume (Dimick et al. 2004; Walker et al. 2013)
Multiple testing	A big issue in monitoring in general: the more units, the greater the number of statistical tests performed and the greater the chance of a false positive result.	Too many false alarms are like the boy who cried "wolf" and will lead to loss of trust in the system.	See Chapter 9
Correlation	Tests commonly assume that data points are independent. Patients treated by the same doctor or hospital are not statistically independent.	The standard error will be wrong (generally too small), potentially leading to too many units being flagged as good and poor performers.	Physician scorecards (Hofer and Hayward 1999; Landon and Normand 2008)
Attribution	Staff often work in teams or networks. To whom should we attribute the outcomes of patients transferred from one colleague or hospital to another?	If there are care quality problems, then improvement cannot happen without identifying where they lie. Being labelled a poor performer, especially wrongly, can be devastating for the individual staff member or unit.	Kosseim et al. (2006) on AMI mortality; CMS publicly reported metrics (Dorsey et al. 2015)
Modifiable areal unit problem	This occurs when conclusions of a study are influenced by the way data are aggregated, in terms of either scale or boundary definition. It's an issue if you want to report by geographical area rather than a fixed entity like a practice or hospital.	"Hotspots" of high disease incidence, for example, can appear or disappear depending purely on your choice of unit.	Arsenault et al. (2013) consider nine criteria for choosing the unit; the ECHO project considers variation by hospital and different types of geographical units

The issues in the table are somewhat interrelated. Let's assume that we have one observation per patient and that we can ignore any clustering that might exist. We now need to aggregate these patient-specific measures up to some institutional level; we'll consider aggregation to some higher geographical level shortly. Sometimes, there's more than one statistically valid choice for reporting; for example, figures for a given specialty could be given by hospital site if the hospital as an organisation (such as NHS Hospital Trust) operates that speciality at several sites. It's clear that the greater the aggregation, the larger the unit and the greater the sample size at each unit, which gives more power to detect differences between units. A key drawback of aggregation is the potential for losing interpretability and attribution. There is a trade-off here; indeed, trade-offs will become a theme throughout the book. Reporting post-operative mortality rates by hospitals will yield more robust estimates of those rates but will hide variation between surgeons within the same hospital that could be important. Reporting by surgeons, pioneered for CABG in New York State in the 1980s and increasingly common, will incur more random error and reduced – indeed, often low – power to detect differences in performance (Dimick et al. 2004; Walker et al. 2013) but is more actionable. There is growing literature on the impact of sample size and physician/hospital report cards. The statistical reliability of performance measures has been estimated – and sometimes found wanting – by the physician (Hofer et al. 1999) and surgeon (Dimick et al. 2012). A technical tutorial on the reliability of unit comparisons was produced by the RAND Corporation (Adams 2009). This issue is considered in more detail in Chapter 9.

The lower the level of aggregation, the more units there are to compare: a few hundred hospitals may employ thousands of surgeons. As with medical tests, statistical tests all have false positive rates. The conventional cut-off p value (also known as the type I error rate) of 0.05 that represents the threshold for statistical significance implies that if a test is performed on a sample drawn randomly from the same underlying population, then up to 5% of those samples will show a "statistically significant" difference with $p < 0.05$ even when none in fact exists. Another way of saying this is that, of a hundred surgeons all with the same post-operative mortality rate, five on average would likely show up as either high or low outliers. Each statistical comparison between a unit and the benchmark such as the national average (the choice of benchmarks will be discussed in Chapter 9) has a false positive rate, so more comparisons means more false positives. The frequency of monitoring shares these issues. Annual comparisons of performance require fewer tests than quarterly or monthly monitoring. They will also be more numerically robust, but they will hide fluctuations during the year such as seasonal patterns. Of course, you also have to wait a year for the results, which is a long time if performance has been poor. Ways to deal with multiple testing are considered in Section 9.4.

This discussion assumes that healthcare units are fixed in space and time. In public health monitoring, however, the unit may be geographical rather than institutional. We may need to monitor disease incidence rates by area. Some healthcare performance measures such as access to treatment can also be useful when calculated by a geographical area rather than by healthcare provider. It is natural and convenient to use existing administrative areas as the basis, given that they correspond to institutions of responsibility and tend to exist as items on available databases. However, infectious disease vectors and countless other pathogens don't recognise neat human boundaries. Even if populations were not mobile and everyone spent their lives entirely within their village, electoral area, district, county or region, these units and their boundaries are artificial constructions and can be periodically redefined, for instance, by electoral commissions. Changing the boundaries is unlikely to influence the causes of disease or the effects of treatment but can, at a stroke, change whether a given area has high or low rates of whatever we are measuring. Our analysis is thereby at risk of the modifiable areal unit problem. In this example, it's the electoral commission who have modified the areas, but we as the performance monitors may be able to change or at least choose them. The ECHO project, which we'll meet again in Chapter 10, compares performance indicators by the geographical area and provider across several European countries. One of the project's aims is to quantify the non-random variation between units. The project leaders recognise that the greater the heterogeneity in the size of the units, the greater the chance of observing extra variation that's attributable to the population's size (noise) rather than to variations in performance (signal). The noise is exaggerated for rare binary outcomes compared with more common ones.

There are several approaches to tackling the modifiable areal unit problem and reduce the influence of chance. The simplest is to aggregate data by combining more years or by merging units. However, this can hide time trends and produce artificial units that are not relevant to stakeholders. People whose identity is partly defined by belonging to "Community X" will not be happy if you lump them together with neighbouring "Community Y" to create the "Community of X and Y", which will mean nothing to residents of either. That won't matter too much if the results are not reported publicly, but the new units might also not reflect local healthcare policy decision making and commissioning. More sophisticated is to use some spatial smoothing technique to tune down the noise, but this runs the risk of smoothing away genuine differences in performance. The ECHO approach for England, Denmark and Portugal was to construct intermediate-sized areas that reflect the actual population "exposure" to hospital care, based on each existing administrative area's use of inpatient and day case care at each hospital. Technical details are given at http://www.echo-health.eu/handbook/unit_analysis.html. These issues are also important in international comparisons, which involve units of enormous heterogeneity, both in size and outcome prevalence, within and across countries (Chapter 10). Choosing the geographical unit of analysis is therefore crucial. Arsenault et al. (2013) considered nine criteria for choosing the unit when looking to detect regions with high or low disease incidence, covering many aspects such as biological relevance, communicability of results, ease of data access, distribution of exposure variables, cases and population, and shape of unit. They give some recommendations, and the interested reader is directed there.

3.1.2 Episode Treatment Groups

Thus far, we have discussed having one or more observations per patient in a way that implies a neat database where each observation has a discrete time-stamp. This works well for surgery, a single date-stamped intervention with a relatively clear demarcation of what's before and what's after the intervention. It can also work well for acute medical events resulting in hospitalisation, where the admission rather than the patient is typically taken as the unit of analysis. For chronic diseases, in particular, however, it's less obvious what to do. While an admission may work well for stroke, it may work less well for heart failure, leukaemia or dementia. An alternative concept is the *episode of care*. This has been defined as "all clinically related services for one patient for a discrete diagnostic condition from the onset of symptoms until treatment is complete". Another definition is "a series of temporally contiguous healthcare services related to the treatment of a given spell of illness or provided in response to a specific request by the patient or other relevant entity" (Hornbrook et al. 1985). Such a concept is more patient-centred and moves away from the narrow focus on an individual provider to look at how multiple providers and the whole team work together to manage the patient.

An episode can belong to only one patient, but a patient can be in multiple episodes at the same time. Episodes are defined and grouped by proprietary software in the United States for purposes such as value-based purchasing initiatives to improve the quality of care while avoiding unnecessary costs. The market-leading method for grouping such episodes is the Optum™ Symmetry® Episode Treatment Group® (ETG) for medical and pharmacy claims data. It adjusts for disease severity and associated costs and has some similarities with diagnosis-related groups such as its ability to create resource-homogeneous groups. One key difference is that it cuts across healthcare sectors, so it covers inpatients, outpatients and ancillary services. An ETG captures medications, diagnostic information (comorbidities and complications) and procedures.

The key features of ETGs are anchor records, clusters and non-anchor (i.e., ancillary and pharmacy) records. Anchor records demonstrate that a clinician has evaluated a patient and determined the types of services required to further identify and treat their condition. An anchor record starts an episode. The software evaluates every ancillary and pharmaceutical service (the non-anchor records) against all episodes to determine the best fit and groups the claims into clusters, each with one anchor record. The episode ends with a second anchor record. Each episode is assigned to a "base ETG", which classifies the medical condition such as diabetes or COPD. The base ETG code has six digits to describe it, with a seventh as a flag for complications, an eight as a flag for comorbidities and a ninth as a flag for treatment types (e.g., surgery or chemotherapy). Inbuilt hierarchies dealt with potential conflicts. A severity score is obtained for each base ETG via multiple linear regression models that determine the impact of the comorbidities, and so on, on costs. Lastly, each base ETG is mapped to one major practice category, which represents a body system or physician specialty. For chronic conditions, which are by their nature open-ended, the software arbitrarily partitions episodes into periods of a year to make analysis practical, though the user can choose other options. The definition of an episode differs in the ETG system from that used by the main competitor, the MEG by Thomson/Reuters (Thomson/Reuters Medstat Medical Episode Grouper software): an analysis for Centers for Medicare & Medicaid Services (CMS) found that chronic condition episodes cover about 65% of the costs in ETG but only 43% of the costs in MEG (MaCurdy et al. 2008) and that a proportion of claims and costs could not be allocated to either type of episode group. About half the episodes lasted only a day. Their report highlights a number of assumptions and limitations with such episodes and the challenges when applying the algorithms to CMS data. The algorithms make the following main assumptions:

- All claims relevant for treating a particular illness incident can be grouped into a distinctive episode of care.
- The medical services comprising any claim belong to only one episode.
- Episodes of care have clearly defined start and end dates.

These were originally developed to assess physician cost-efficiency and are in widespread use in the United States. However, they have a number of limitations. As the episode concept makes use of treatment data, and treatment relates largely to physician decision making, ETGs can't be used to assess quality of care in dimensions other than cost. Another limitation of this type of formulation of an episode is that a provider can appear efficient when they perform a lot of procedures that are each at comparatively low cost, even if many are unnecessary and raise the risk of harm. There is uncertainty over whether surgical complications should be grouped with the original surgery or considered separately; likewise for flare-ups of chronic conditions. The patient might not see the same physician for both the index and the later event. This is partly a problem of attribution as more than one physician or healthcare unit in general will often be included in a single episode with the algorithm. Options include taking the unit with the biggest share of the costs or with the most patient contacts; we consider attribution in the next section.

In view of the limitations of an episode as defined thus far, such as the difficulties in handling patients with chronic or multiple chronic conditions and the inability to capture appropriateness or patient preference, the National Quality Forum (NQF) in the United States defined a "generic episode of care" for different kinds of disease, both acute and chronic. This divides the patient's journey into three parts:

1. Population at risk, for whom primary and secondary prevention is important to capture
2. Evaluation and initial management following the onset of clinical illness
3. Follow-up care

The NQF project committee suggested that follow-up care should at least capture outcomes at one year for chronic conditions (NQF 2009). Each phase would have its own set of quality measures (see Chapter 4, where we also consider the particular challenge of assessing the quality of care for people with multiple conditions). The focus of their report was on comparing the efficiency of providers, but the approach also applies to other dimensions of quality. Episodes begin with the onset of symptoms and end at the end of follow-up. Measures of quality should then be at the patient level. The report illustrates the generic episode of care only for AMI (where the episode starts with chest pain and ends 30 days after hospital discharge) and lower back pain (where the episode starts with the pain onset and ends 6 months after the final treatment) and so does not yet provide any algorithms like the ETG system.

3.2 Issues More Relevant to Reporting: Attributing Performance to a Given Unit in a System

Information collected at the level of the patient or, as we have seen with different specialties, at the level of body part needs to be reported at the level of some healthcare unit. Issues particular to public reporting are considered in Chapter 11, so here we concern ourselves with reporting in general. This may be to the hospital board, the regulator or other responsible parties.

We have already touched upon the tricky problem of attribution when comparing surgeon-level and hospital-level reporting of surgical outcomes. Reporting by surgeon rather implies that the surgeon alone is responsible for his or her outcomes. While they may be responsible for their patients' care, they don't of course work in isolation. Nursing, physiotherapy and many other staff have key roles and can affect outcomes. There is, however, generally very little or no information in most databases on the activities of the other staff. Likewise, in primary care the general practitioner (GP) or family doctor is responsible for the patient's care, but primary care electronic databases rarely capture the activity of community health staff. Sometimes, the level of aggregation is limited by the data. The United Kingdom's Quality and Outcomes Framework, a national pay-for-performance scheme running since 2004, involves reporting quality measures by general practice. Many practices, though, employ a number of GPs. Sometimes, however, available databases do capture something of the multidisciplinary nature of healthcare. Hospital databases such as Hospital Episodes Statistics in England and similar systems in the rest of the United Kingdom allow the tracking of patients as they are transferred from one consultant (senior doctor) to another and from one hospital to another. Many countries can link primary care activity to secondary care activity at least locally, if not nationally. If a patient has an operation at hospital H1, is transferred and dies at hospital H2, should our post-operative mortality monitoring system assign that death to H1 where the operation was done or to H2 where the death occurred? If a patient admitted with acute myocardial infarction to H1 is transferred and dies at hospital H2, should our AMI mortality monitoring system assign that death to H1, H2 or both? Transfers are common enough to affect the resulting death rates (Kosseim et al. 2006; Kahn et al. 2007; Bottle et al. 2013b). In the Centers for Medicare & Medicaid Services' National Quality Forum–endorsed AMI mortality measure (see AHRQ http://www.qualitymeasures.ahrq.gov/content.aspx?id=48168), patients who are transferred from another acute care or veterans hospital are excluded because the death is attributed to H1. In our England-wide system, we attribute deaths to both H1 and H2, though the death is only counted once in the risk-adjustment modelling (Bottle and Aylin 2008a). For the official Summary Hospital-wide Mortality Indicator in England's NHS, deaths are attributed to the last acute hospital trust in the admission. Another possible approach is to "divide" the death between H1 and H2 in proportions determined by the time spent in each. While intuitive, this ignores the question of whether the transfer was made at the optimal time and indeed whether it should have been made at all. This difficult issue can be circumvented for the most part by reporting at a higher level of aggregation such as hospital network if it can be said to exist and if it represents a clinically meaningful entity for the set of patients of interest. If that network is deemed to have unacceptable performance, then the figures will need to be disaggregated in order to investigate the causes.

Depending on the performance measure, there are huge differences in how much the health system or a given unit can affect things (Hauck et al. 2003). Waiting times will be more amenable than population mortality – we'll return to performance measures of the whole system in Section 4.5. It might not be so apparent where the main responsibility lies. For instance, if a department in a publicly funded hospital is not performing well, this could be due to inefficient working practices (the responsibility of the hospital management), shortages of resources (the responsibility of the relevant funding body) or inappropriate national wage scales (probably the responsibility of central government) (Hurst and Jee-Hughes 2001). The degree of influence on the chosen performance measures and which are the main levers to pull to improve things need to be considered when holding units to account.

In summary, there are a number of issues to consider when choosing the unit of analysis and reporting. It may be that the choice is restricted by data limitations or political considerations, but we should at least appreciate the implications.

4

What to Measure: Choosing and Defining Indicators

Aims of This Chapter

- To discuss what quality of care means, including some common classifications
- To set out the steps in the creation and validation of a quality indicator

The number of indicators of quality of care in existence is already vast, but not all are created equal. There are also still significant gaps. The choice – whether from among existing ones or purpose-built new ones – must primarily reflect the monitoring goals. Other key considerations are how and by whom the information will be used in practice. In this chapter, we'll consider the following:

- How quality can be defined
- Common indicator taxonomies
- Features of an ideal indicator
- Steps in construction and common issues in definition
- Validation of indicators
- Some strategies for choosing among candidates
- When to withdraw indicators

Along the way, we'll give examples to illustrate these points and how indicators can fall short of the ideal. There will be only brief mention of issues that will be covered in later chapters, such as risk adjustment and presentation of results.

4.1 How Can We Define Quality?

This is remarkably difficult. One definition is "the degree to which health services for individuals and populations increase the likelihood of desired health outcomes and are consistent with current professional knowledge" (Lohr 1990). There are several elements to this. One is the mention of both individuals and populations. With finite resources, there is clearly a balance to be struck between these two. The second point is in the phrase "desired health outcomes". This is easier to agree on, but there can be conflict between what the patient and staff see as desired and what is possible, which relates to the last three words in the quote. What healthcare can do now is more than what it could do last year but less than it will be able to do next year. Third, the word "likelihood" implies that, even with the best care, the desired health outcomes might not be achieved. We are simply increasing the odds as much as we can in the patient's favour. King Hammurabi, who we met in Chapter 2, with his all-or-nothing, success-or-failure physician rating, would not have agreed.

An older definition, however, points out that quality is also a matter of value judgments that are applied to the various dimensions of care (Lee and Jones 1933). These values come not only from the medical care system but also from the society of which that system is a part. In the era of patients as customers, it has strong elements of customer service: the speed at which we get to see a doctor and the manner in which we are treated. If we have to hand over money at the time of the consultation, then the fee and value for money will likely seem more important than if the cost is prepaid via insurance or taxation. If someone else is picking up the tab, then they'll be thinking about the fee and value for money, whereas we'll be more interested in whether our needs were met. Good care is what we say it is. It also depends on who "we" are. Just as beauty is in the eye of the beholder, so it is with good healthcare, and the point of view is a recurring theme in this book. For excellent discussions of what good care might consist of, see, for instance, Donabedian (1966). The rest of this chapter concerns attempts to produce indicators to measure it.

4.2 Common Indicator Taxonomies

The most influential classification divides indicators into structure, process and outcome and was proposed by Donabedian (1966). He first set out the issues regarding how quality of care could be defined before discussing how the environment in which care occurs (assessed by measures of structure) and whether "medicine is properly practiced" (assessed by measures of process) affect outcomes:

$$\text{structure} + \text{process} = \text{outcomes}$$

This gives us three types of measure, each with its strengths and limitations. There are various good reviews of these (Rubin et al. 2001; Mainz 2003a,b); Table 4.1 summarises their features, pros and cons. In short, outcome indicators are best for monitoring progress on performance goals, and structure and process indicators are best as actionable levers to drive improvement towards those performance goals (Raleigh 2012).

Outcomes may be divided into intermediate and end result. For someone with diabetes, a high HbA1c is an intermediate step on the way to several more serious outcomes later. Mortality is clearly in the end result camp. Other outcomes such as renal failure and lower limb amputation would generally be regarded as end results as well, as they do not necessarily lead to death, though they increase the risk of it. To be useful, intermediate outcomes should have an evidence base that relates them to end result outcomes. They need to be good proxies and/or of interest in their own right.

Another influential classification divides the concept of quality into domains. The Institute of Medicine's analytical framework (Corrigan et al. 2001) uses six domains, which are defined as follows:

1. *Safe*: Avoiding harm to patients from the care that is intended to help them
2. *Effective*: Providing services based on scientific knowledge to all who could benefit and refraining from providing services to those not likely to benefit (avoiding underuse and misuse, respectively)
3. *Patient-centred*: Providing care that is respectful of and responsive to individual patient preferences, needs and values and ensuring that patient values guide all clinical decisions
4. *Timely*: Reducing waits and sometimes harmful delays for both those who receive and those who give care
5. *Efficient*: Avoiding waste, including waste of equipment, supplies, ideas and energy
6. *Equitable*: Providing care that does not vary in quality because of personal characteristics such as gender, ethnicity, geographic location and socioeconomic status

England's national regulator, the Care Quality Commission, assesses units on these domains, which have a lot of overlap with the IoM's:

1. Safe (avoiding harm)
2. Effective (delivered in line with national guidelines)

TABLE 4.1

Structure, Process and Outcome Indicators Compared

Feature	Structure	Process	Outcome
Relation with quality	Can be hard to assess	Direct measure of quality if evidence based	Variable, as the patient may have poor outcome despite good care On the other hand, can reflect sum of all processes of care that the patient had
Attribution	Relatively straightforward to assign	Relatively straightforward to assign	May be difficult to attribute to a single institution or process, e.g., hospital death, if more than one institution involved in care
Evidence base for relation with quality	Often exists	Most evidence-based indicators are process related Relation with outcomes may be weak	Often used as the gold standard when assessing new candidate process or structure measure
Data availability and cost	Often good and low cost	Easy to measure but many not available in routine data	Hard to define for some specialties, e.g., mental health Some important ones in routine data Time lag between care and outcome can be long
Actionability	Can be limited, e.g., a small hospital can't easily become a large one	High	Low (directly) and attribution not always clear: need to follow up by examining processes
Existence of measures	Limited, especially for ambulatory care	Many	Many, though more for surgical than medical specialties
Relevance to patients	Limited	Limited	High
Risk of gaming	Low	Highest of the three types Can become a tick-box exercise	Low, but higher for some information used in risk adjustment. Concerns over potential for avoiding high-risk patients
Statistical issues	Few, as measure can be as simple as unit has MRI scanner, yes/no?	Usually higher power than outcomes Number of processes can be large -> multiple testing or need for composite Ceiling effects, when everyone performs very well, limit usefulness	Commonly used binary outcomes can be rare -> low power Risk adjustment usually needed, which is tricky

3. Caring (direct staff–patient interactions are caring)
4. Responsive (meets people's needs such as access, comfort)
5. Well led (at hospital, service and ward/team level)

Effectiveness and safety have thus far been addressed by developers far more than the other domains. Regarding equity, it should be noted that fair, needs-based access only applies to good-quality, evidence-based care. Access to services is not enough in itself if those services are no better than snake oil. Equity also implies fairness, and that includes fairness in financing and financial risk protection for households: see Murray and Frenk (2000) for a discussion of fair financing for the healthcare system in different political contexts. We will consider efficiency in Section 4.4 under whole-system measures.

There is also concern that some terminally ill patients are receiving "heroic" interventions that, despite being done with good intentions, cause unnecessary distress as well as expense. The concept of patient-centredness echoes the 1933 definition of quality mentioned earlier that talks of medical and particularly societal values but here focused on the individual. One person's respectful is another person's insulting. The other dimensions implicitly reflect our values too; for example, a different set of values could make equity irrelevant if we as a society decided that it's acceptable for quality to vary according to our ability to pay for it. Patient-centredness is typically assessed using surveys, and the Picker Institute is among the best-known purveyors. More broadly, the most commonly researched approaches for measuring patient and carer experience, of which patient-centredness is a key element, include surveys, interviews and patient stories. Launched in April 2013, the Friends and Family Test simply asks NHS patients if they would recommend the services they have used. The results are published monthly on both the NHS England and NHS Choices websites. For more on PREMs, see, for example, the Health Foundation report on patient experience tools (de Silva 2013). Analogous to PREMs are PROMs or PROs (patient-reported outcome measures or outcomes), collected by asking the patient for their views on how their treatment has affected their condition and/or their life. These may be generic, such as the EQ-5D™ with its five domains of mobility, self-care, ability to perform usual activities, pain/discomfort, and anxiety/depression, or they can be condition-specific, such as the Oxford hip and knee scores for joint replacement surgery. These kinds of measures are better at describing how effective the treatment was for the patient than the cost or the complication rate.

There are a few other useful classifications. One divides indicators, particularly outcome measures, into rate-based or sentinel. In the former, there is some background "acceptable" level of undesirable outcomes that current medical practice is unable to reduce. In the latter, however, the adverse outcome should have a rate of zero. These are also known as "never events", such as wrong-site surgery. An investigation should follow even a single occurrence of these. Other classifications include grouping by function (e.g., screening, diagnosis, treatment and follow-up), type of care (preventive, acute, or chronic) and whether the indicator is disease-specific or "generic".

Domains like those listed earlier from the IoM or the CQC, commonly form the elements of an overarching measurement *framework* that largely dictates the indicators that are subsequently chosen. Many commentators have argued for the need for a robust conceptual framework that covers all the major domains of the healthcare system and that aligns with the system's broader strategic goals and priorities. For example, in Australia since 2011, the national Performance and Accountability Framework aims to "support a safe, high-quality Australian health system through improved transparency and accountability" and underpins reporting across three domains: equity, effectiveness and efficiency of service delivery; for details, see the National Health Performance Authority website. The performance measurement system should also be aligned with other policy and financial priorities. By setting out a clearly defined vision and by involving potential end users, it can encourage buy-in from clinicians and consumers and can ensure its future usefulness. Smith et al. (2009, p. 678) argue that an optimal framework would also fit a health system's existing organisational structure and accountability arrangements. The alternative to this is continued fragmentation and underperformance in some parts of the system.

4.3 Particular Challenges of Measuring Patient Safety

Patient safety can be defined as "the avoidance, prevention and amelioration of adverse outcomes or injuries stemming from the process of healthcare" (Vincent 2010; Vincent et al. 2013). It's generally about failure, whereas the other aspects of quality of care relate to the intended results of healthcare. Some process measures can be considered to cover both quality and safety. Failure to correctly implement a process that leads directly to harm is likely to be viewed as a safety issue: the spinal administration of vincristine causing death is an extreme example of this. In contrast, not giving beta blockers after discharge from hospital following a heart attack is more likely to be considered a quality rather than a safety issue even if the patient then has another heart attack. The reason is that the link with the second acute myocardial infarction (AMI) is less clear-cut than between intrathecal vincristine and death.

After some landmark studies put alarming estimates on the scale of not just error but actual harm caused by medical care in hospitals, improving patient safety has become a global movement (Brennan et al. 1991; Wilson et al. 1995). However, even more so than for other aspects of quality, the obstacles to improvement are considerable. They include the need for workable definitions and clear conceptual frameworks for measurement, the existence of multiple perspectives and the fact that, as found in other industries, catastrophic errors are not always foreseen or predicted by the safety data that are collected. What is considered a safety rather than a quality issue changes over time, such as with healthcare-associated infections, in-hospital falls and pressure ulcers. The debate over the relative merits of focusing on error or harm is akin to that over process or outcome. Harm (the outcome) is what patients care most about. We'll all put up with errors in our care (processes) at least to some extent as long as we don't come to harm. Many errors are of no practical consequence for the patient and can facilitate learning for staff.

There are different types of harm. A number of classifications exist, but harms can be treatment specific (e.g., adverse drug reactions, surgical complications), due to overtreatment (accidental or deliberate), general (e.g., falls and dehydration, though some patients are clearly at higher risk of these than others), due to failure to provide appropriate treatment (either at all or early enough), due to delayed or inadequate diagnosis, and, finally, psychological. Incident reporting systems (see Section 5.5) also tend to capture a lot of near misses – an error and potential harm nearly happened but were prevented by the actions of staff (either the person who was about to make the error or a colleague), patients and/or carers or relatives. It should be noted that even near misses can cause psychological harm to the patient and their relatives. The World Alliance for Patient Safety of the World Health Organization (WHO) recognised healthcare's inability to classify, aggregate and compare patient safety information across organisations internationally. The WHO, therefore, initiated the International Classification for Patient Safety (IC4PS) in 2005. Many other classifications of the severity of harm exist. For example, the National Reporting and Learning System (NRLS) was rolled out in 2003 across England and Wales to capture incidents and near misses and grades incidents from no harm, through low, moderate and severe, and, finally, death. In the United States, the National Coordinating Council for Medication Error Reporting and Prevention produced the country's first comprehensive taxonomy for studying medication errors. Its index comprises categories A to I: potential for error (category A), actual error that did not reach the patient (category B), actual error that reached the patient but did not result in harm (categories C or D) and actual error that reached the patient and resulted in harm (categories E, F, G, H and I), where E is "an error occurred that may have contributed to or resulted in temporary harm to the patient and required intervention" and I is death ("an error occurred that may have contributed to or resulted in the patient's death"). For such classifications to be useful, how severe the harm is rated should not depend on who is rating it. Inter-rater reliability has been assessed as moderate (kappa = 0.61, but notably higher if only six categories were used) (Forrey et al. 2007).

Safety is one dimension of quality but is itself multidimensional. The focus is moving from counting errors or harms after the event towards looking at hazards that might lead to error before the harm occurs. The report published by the Health Foundation in 2013 (Vincent et al. 2013) put forward five fundamental classes of safety information to reflect the multidimensional nature:

1. Measures of harm (in different forms)
2. Measures of reliability, which encompass behaviour, processes and systems
3. The information and capacity to monitor safety hourly or daily
4. The ability to anticipate problems and be prepared
5. Integration of and learning from safety information

While a safe system avoids harm, another prerequisite is that its basic processes are very reliable. The concept of the high-reliability organisation has its origins in the 1980s work by the Berkeley group on U.S. naval aircraft carrier operations, air traffic control and nuclear power. Harm and reliability have had some attention from methodologists, though we're far from able to say whether healthcare is becoming safer (Vincent et al. 2008). The Health Foundation's case studies found that information on problem anticipation and preparedness was by far the hardest of the five classes to produce. Their report concludes with ten key points for safety measurement and monitoring. The first is that a single measure of safety is a fantasy; given that safety is a dimension of quality, the same can be said regarding quality. Another key point is that in the day-to-day monitoring of safety, anticipation and preparedness must be customised to local settings and circumstances as these are local activities. That does not imply a measurement free-for-all, and nor does it preclude standardised national or international metrics. Empowering clinical units to adapt measures for their specific context helps with buy-in and morale. There is also a need to adapt safety metrics developed for the acute care sector to other sectors such as primary care and mental health.

Organisational culture has received a lot of attention in the literature. Safety culture stems from the attitudes and values of the staff and has been assessed using safety culture maturity matrices. These display a set of key indicators on the y-axis and an evolutionary measure of cultural maturity on the x-axis. Originally developed as part of the "Hearts and Minds" project for Shell plc, one such matrix, the Manchester Patient Safety Framework (MaPSaF), was adapted for use in primary care and has since been developed into versions for other settings (Ashcroft et al. 2005). The MaPSaF describes five levels of increasingly mature organisational safety culture, from "pathological" (least mature) to "generative" (most mature). Information is gathered by setting up a series of focus groups where healthcare staff assess the maturity of their local team and organisation. Staff attitudes to safety and culture can also be obtained from surveys (see Chapter 6).

Some other elements of safety relate to working practices. Staff handovers, such as from one shift to the next, teamwork and continuity of care can all affect outcomes. Many safety indicators are of particular interest and relevance to nurses. In a position statement by the UK Royal College of Nursing (RCN 2009), the priorities for indicator development were given as: high-risk/high-cost topics, particularly pressure ulcers, failure to rescue (death following treatment complications) and falls; the essence of care; patient-reported outcomes, such as patient experience and perception of patient involvement; and systems of care elements such as continuity of care, teamwork and staffing levels.

Readers seeking an in-depth account of all aspects of patient safety including, but certainly not limited to, its measurement should look no further than Charles Vincent's book on the subject (Vincent 2010).

4.4 Particular Challenges of Multimorbidity

Many people, including over a quarter of the U.S. population, have multiple chronic conditions (MCCs), and the proportion is rising around the world. They are at higher risk of poor health and poor care than those with fewer conditions, yet they are not well served by existing quality measures. Clinical guidelines typically focus on the management of a single disease. A concern in people with MCCs is potential overtreatment instead of treatment that's appropriate for the individual. Close adherence to current indicators that are relevant to single diseases could actually result in harm in those with MCCs (Lee and Walter 2011). By their nature, chronic conditions change over time, as do the patient's treatment goals, life goals and also their preferences (see later text). Assessing whether the goals and preferences are met at any given point in time will need a lot of data and some way of summarising them.

The National Quality Forum (NQF) convened a multi-stakeholder steering committee to develop a measurement framework for this growing population segment. The first task was to define MCC. One option is to take a simple count of concurrent diseases and flag people with two or more. This doesn't consider the complexity or interaction among the conditions. Two related terms that go beyond the simple count is "complex patient" and "complex elderly". Patient complexity recognises the social or non-medical issues that are not easily amenable to healthcare intervention. However, while the term complexity considers how the interactions between conditions impact on the need for care, they don't capture functional status or quality of life. It's trickier still with children because of the types of chronic conditions they experience (largely due to genetic conditions, the consequences of low birth weight and prematurity, and environmental factors) and because the impact changes as the children grow and mature. Children with MCCs are particularly badly served by available quality measures. After consideration, the NQF's committee defined people with MCCs as having "two or more concurrent chronic conditions that collectively have an adverse effect on health status, function, or quality of life and that require complex healthcare management, decision-making, or coordination" (NQF 2012).

In terms of what to measure, the NQF report set out some guiding principles. These included assessing the success of the coordination of care among the various providers; trying to identify whether treatment is appropriate (often described via measures of overuse, underuse and misuse); incorporating patient and carer preferences; monitoring over time, which would involve demonstrating improvement or maintenance; where possible, not excluding patients from quality measures; risk-adjustment outcomes with caution so as not to hide care gaps (see Chapter 6); and taking data in a standardised way from difference sources, as "no one data source is adequate" (see Chapter 5). To try to limit the measurement burden, the NQF recommends prioritising measures with good links to patient outcomes and patient preferences.

We have already discussed the differences between process and outcome measures, but a concept that is particularly relevant to MCCs is continuity of care. It's defined as consistent, "seamless" treatment over time involving various healthcare providers and settings. It also refers to long-term care by a healthcare team that includes effective communication for information sharing. Assessing this has received considerable attention. A 2006 systematic review (Lee and Cabana 2006) identified 32 different indices used to measure continuity of care: 17 on the density of visits, 8 on the dispersion of providers (how the health service contacts are spread among the providers), 4 on subjective estimates, 2 primarily based on the duration of the relationship with the provider and 1 on the sequence of providers. None considered the type of visit. The usual provider continuity (or usual provider of care [UPC]) index is commonly used in primary care and is the proportion of consultations which were with the GP who the patient saw most often. It's calculated for each patient and then averaged across all the patients in the sample, for example, a two-year period. A drawback is that it includes patients who have had just one appointment in total and so will inevitably have a UPC of 100%, which will push up the overall average. As an alternative, the popular modified modified continuity index (MMCI) focuses on the "dispersion" between providers and is based on the number of providers and number of visits only. Index values range from 0 (each visit made to a different physician) to 1 (all visits made to a single physician). More recently, the literature has focused on issues such as care by the smallest number of professionals and the continuity of aspects such as information, consistency of clinical management among the teams and the ongoing relations between patient and staff. Continuity of care could be said to have three main elements: seeing the same staff member over time, maintaining a good patient–professional relationship and having coordinated care. These are related but different, and various commentators have argued that they should be measured separately.

See, for instance, Dreiher et al. (2012) for a review of measures and relevant literature.

Harder to measure is patient preference. There are two aspects to this: preference for a given health state (this is known as utility) and for a given treatment, and the two are not necessarily linked. For instance, a patient might wish to avoid major surgery even if it increases their chances of survival. A lot of methodological and applied work has been done on the association between patients' preferences and treatment choices, though this is outside our scope. Much of the research on patients' preferences derives population-based utility values of quality of life in relation to different health end states using standard gamble or time trade-off techniques. With the latter, patients are given information in hypothetical scenarios about the treatment options, outcomes and probabilities. With one combination of these, the patient might choose one particular treatment. The efficacy of the treatment is then varied by the researcher until patients decide they want the alternative one. The method for estimating these preferences can affect the results. Approaches other than this kind of trade-off have been used: see, for example, the introductory article by Bowling and Ebrahim (2001) and the supplementary issue of the journal in which their article appears.

There is another angle to patient preference that was mentioned earlier in this section in passing, which is certainly worth a little more space before we move on: choosing the goals of treatment. There is a growing movement to focus on measuring the outcomes that matter to patients. For example, clinical trials of medications for mental health conditions might have treatment compliance or discontinuation as their primary endpoint on which the new medication will be judged. Treatment compliance in itself does not matter one bit to most patients unless the reason they stopped taking the medication was side effects or other inconveniences such as the need for frequent monitoring of blood levels. This does *not* mean that one should *not* monitor compliance, just that we mustn't lose sight of what's of most importance.

4.5 Measuring the Health of the Population and Quality of the Whole Healthcare System

A common limitation is when indicators relate to a fairly small part of the healthcare provider or a small part of the patient's journey through their diagnosis and treatment, such as in-hospital mortality after surgery for cancer. We also need indicators for the whole healthcare system. More generally, we also need to monitor the health of the whole population and some large subgroups of it such as infants and people with chronic conditions. Before we give some examples of both these types of indicators and end with a look at the WHO's approach, we need to tackle two issues regarding the healthcare system: what are its boundaries, and what proportion of the people's health can we fairly attribute to its quality?

Defining the boundaries of a healthcare system beyond the physical walls of a hospital, for instance, is hard. For example, the IHI developed and piloted a set of 13 Whole System Measures to measure the overall quality of a healthcare system (Martin et al. 2007). They define the "system" as "an integrated health system, a multiple hospital system, a free-standing hospital, or an ambulatory care organization". A number of these measures such as overall hospital mortality and readmission rates are hospital-focused, and we need to go further. Murray and Frenk (2000) began their definition of the system with the concept of a "health action". This is any set of activities whose "primary intent is to improve or maintain health". The healthcare system would then include the resources, actors, and institutions related to the financing, regulation and provision of health actions. This makes it very broad, though it still doesn't include everything that has an influence on health, such as education, if its primary intention isn't to improve health. That doesn't invalidate the definition, though. The effect can go the other way too as our health can affect our progress in education. Health, the healthcare system and how it's financed can also affect the economy (e.g., employment-based insurance can reduce labour mobility), though the interrelationships and necessary data are too complex to be monitored routinely. Following on from this, some people distinguish between the "healthcare system" and the "health system", with the latter covering anything that impacts or determines health in its broadest sense within a given society; see Smith et al. (2009) for more on this distinction and on the degree to which healthcare influences health. It's clear that international comparisons can only be fair if they use consistent definitions of each system.

Countries need to monitor their population's health both in terms of actual disease and key determinants or risk factors of disease. Common examples of the latter include tobacco use, obesity, air and water quality, housing, employment, poverty and diet. These are only partly within the direct control and indeed responsibility of the healthcare system, so should that system be accountable for the levels of these factors? Murray and Frenk (2000) argue that they should. They qualify this by saying that performance of the health system should be relative to what could be achieved for that country's level of resources. Rich countries will have healthier populations than poor countries, but could they be doing better for their financial input? Are there equally rich – or equally poor – countries who have achieved greater reductions in their rates of smoking, for instance? Most countries are engaged with this bang-per-buck question.

In terms of measuring the health of the population and quality of the whole healthcare system, the WHO and Organisation for Economic Cooperation and Development (OECD) have naturally been active in this area. A 2001 OECD report (Hurst and Jee-Hughes 2001) compared indicators put forward by its member states with those of the WHO, focusing on four countries because of the length of their track record or scale of recent effort: Australia, Canada, the United Kingdom and the United States. Indicators came in four categories: health outcome, responsiveness, equity and efficiency. There were attempts with some indicators, particularly the outcomes, to make them more specific to healthcare in terms of attribution. We have already seen that mortality and other outcomes are influenced by a large range of factors, many beyond the control of healthcare. Low birth weight is commonly used, though the use of prenatal monitoring – an action that is specific to the healthcare system – is just one factor that affects it. This has led to the development of indicators of "avoidable mortality" or "mortality in conditions amenable to healthcare". For monitoring population health, however, the simpler overall or cause-specific mortality rates are sufficient. The second of the four indicator categories, responsiveness, related to how the system interacts with consumers, so it covers patient satisfaction, patient acceptability and patient experience. Third, equity covered various aspects of quality such as outcomes, access, responsiveness and finance and can be assessed for many population subgroups as we've already seen. Lastly, OECD member countries suggested efficiency indicators such as unit costs and length of hospital stay.

In 2008, Australian health ministers met to develop a performance indicator set by which the nation could judge the performance of the system as a whole (Australian Institute of Health and Welfare 2008). Following stakeholder consultation, they took into account the National Health Performance Framework and planned system reforms including the data they already had and indicators already in use. They acknowledged that for many indicators, responsibilities overlapped among both the Commonwealth and the state and territory governments. Some were disease-specific and some sector-specific. They proposed 40 indicators under the headings "better health" (e.g., life expectancy, rate of preventable diseases and injury), "focus on prevention" (e.g., proportion of children who had had all their developmental health checks, cancer screening rates), "access" (e.g., waiting times, out-of-pocket costs), "high quality – appropriate" and "high quality – safe" (e.g., antenatal visits, adverse events), "integration and continuity of care" (e.g., discharge plans for complex care needs), "patient-centred" (patient experience), "efficiency/value for money" (e.g., cost per casemix-adjusted acute hospital admission) and "sustainable" (e.g., workforce changes, proportion of health expenditure spent on health research and development).

It's worth mentioning out-of-pocket costs in a bit more detail before moving on. In rich countries, we are used to insurance co-payments, drug prescription charges and hospital parking fees, but in poor countries the direct costs of healthcare to the individual person and their household can be catastrophic. Good health systems protect people not just as best they can from illness and death but also from the financial consequences of illness and death, or at least the financial consequences associated with medical care. There are various methods for measuring the system's ability to protect people from these consequences. Many studies simply examine the distribution of catastrophic health expenditures, defined as health spending that exceeds a threshold that's usually defined in relation to the household's prepayment income. Ideally, costs should capture not just those of receiving medical care but also income losses associated with illness, injury and death. For more on this, see the chapter by Wagstaff in Smith et al. (2009).

As these examples suggest, measurement of the quality of the "whole healthcare system" is mostly done by breaking it down into segments, both of the population and of the system. In the next section, we'll look at the WHO's approach, which includes some measures such as years of life lost that can be argued to truly encapsulate the performance of the whole system.

4.5.1 The WHO Annual World Health Statistics Report

Specifically, we'll consider here the 2014 WHO World Health Statistics report (World Health Organization 2014). Their first report was published in 2000 amid enormous media and expert attention and assessed the health system performance in 191 member countries. The ranking threw up some surprises. The assessment of Spain as having the seventh-best healthcare system in the world was used by its under-fire government as defence against demonstrators protesting about long waiting times and short primary care consultation times. The report defined health systems and set out why they matter, their evolution, their goals and their functions before proposing a performance framework for member countries. It laid out three goals of a health system:

1. Improve health, both its average level and its distribution (i.e., reducing inequalities)
2. Respond to the expectations of the population
3. Assure fairness of the distribution of financial contribution

Since 2000, the WHO has produced annual reports on how well each country has met these goals. The WHO defines the five major components of health system performance as the overall level of population health, the distribution of health in the population, the overall level of responsiveness, the distribution of responsiveness within the population and the distribution of the health system's financial burden within the population. Performance is estimated from the weighted sum of these five components, and an "overall indicator of attainment" is calculated. Performance is thereby a composite measure (see Chapter 8), with weights taken from a survey of 1000 "key informants" from 125 countries; weighting low mortality higher than responsiveness could have boosted Spain's ranking. These annual reports are available as pdfs from 2005 onwards on their Global Health Observatory website. For a critique of the limitations of the inaugural report, see Navarro (2000).

BOX 4.1   HEALTH-RELATED MILLENNIUM DEVELOPMENT GOAL INDICATORS

1. Under-five mortality rate
2. Measles immunisation coverage among one year olds (%)
3. Maternal mortality ratio
4. Births attended by skilled health personnel (%)
5. Antenatal care coverage (%): at least one visit and at least four visits
6. Unmet need for family planning (%)
7. HIV prevalence
8. Antiretroviral therapy coverage among people eligible for treatment (%)
9. Children aged <5 years sleeping under insecticide-treated nets (%)
10. Children aged <5 years with fever who received treatment with any antimalarial (%)
11. Tuberculosis mortality rate
12. Proportion of population without access to improved drinking-water sources
13. Proportion of population without access to improved sanitation

Data for the 2014 report came from databases from WHO technical programs and regional offices, with extra demographic and socioeconomic data coming from various United Nations organisations and the World Bank. Indicators were included depending on their relevance to global public health, the availability and quality of the data, and the reliability and comparability of the results. The report began with an update on the progress of the 13 health-related Millennium Development Goals (Box 4.1). Several were displayed as average annual rates of decline, though the box just gives the indicators.

The second part focused on a few specific topics such as maternal mortality and child obesity. The third part described trends and compared countries for a large number of global health indicators under the following nine headings:

1. Life expectancy and mortality
2. Cause-specific mortality and morbidity
3. Selected infectious diseases
4. Health service coverage
5. Risk factors
6. Health systems
7. Health expenditure
8. Health inequities
9. Demographic and socioeconomic statistics

The health systems set would fit with Donabedian's structural group of indicators and counts different types of staff and equipment per head of population. The health service coverage and inequities sets are dominated by maternity and child health. It's worth distinguishing here between inequality and inequity, which have some similarities but are not interchangeable. Inequality simply refers to the uneven distribution of health or health resources due to factors such as chance, biology and personal health and/or lifestyle decisions. It's usually used as a general, descriptive term without necessarily any implied moral judgment. In contrast, inequity implies unfairness in this uneven distribution so that it leads to differences in avoidable diseases and deaths that have arisen because of corruption, failure of governance or cultural prejudice. For more on this, see, for instance, the discussion by Kawachi et al. (2002).

Mortality appears in several sets and is measured in different ways (Box 4.2). The measures in Box 4.2 all use all-cause mortality and thereby avoid potential issues with an inconsistent assignation of the cause of death. Observed differences in cause-specific death rates are due in part to differences in diagnostic patterns, death certification or cause of death coding in each country. Even for all-cause mortality, a suite of angles to view this key outcome is needed. The usefulness of knowing the average life expectancy at birth is reduced when neonatal, infant or child deaths are common. After childhood, chronic diseases take over in importance from infectious ones, and the prevalence and treatment of them will be reflected in the adult death rates. The years of life lost (YLL) due to premature mortality represent

---

BOX 4.2 MORTALITY-RELATED INDICATORS USED IN THE 2014 WHO WORLD HEALTH STATISTICS REPORT

- Life expectancy at birth (years)
- Life expectancy at age 60 (years)
- Neonatal mortality rate (per 1000 live births)
- Infant mortality rate (probability of dying by age 1 per 1000 live births)
- Under-five mortality rate (probability of dying by age 5 per 1000 live births)
- Adult mortality rate (probability of dying between 15 and 60 years of age per 1000 population)
- Years of life lost (per 100,000 population)
- Number of deaths among children aged <5 years (000s)
- Maternal mortality ratio (per 100,000 live births)
- Under-five mortality rate (probability of dying by age 5 per 1000 live births)

a different way of summarising the impact of death on a population. They are calculated from the number of deaths at each age multiplied by a global standard life expectancy of the age at which death occurs. The WHO used a standard life table with a projected 2050 life expectancy at birth of 92 years. This is meant to represent a person's potential maximum life expectancy at a given age and is used for both males and females. A death at birth will result in 92.0 YLL and a death at age 70 in 23.2 YLL. These can be split by cause too. The top three causes of YLL in 2012 were ischaemic heart disease, lower respiratory infections (such as pneumonia) and stroke.

Related to the concept of YLL are disability-adjusted life-years (DALYs). One DALY can be thought of as one lost year of "healthy" life, where healthy means free of disease and disability. DALYs are the sum of years lost to premature mortality (YLL) and years lost to disability (YLD). The latter can be based on either new cases (incidence) or existing cases (prevalence), with prevalence used for the 2010 Global Burden of Disease Study. It also involves a disability weight that reflects the severity of the disease on a scale from 0 (perfect health) to 1 (dead).

The flip side of DALYs can be estimated by the healthy life expectancy (HALE) at birth, which the WHO also gives in its report. As the WHO's web page explains, this measure "adds up expectation of life for different health states, adjusted for severity distribution, making it sensitive to changes over time or differences between countries in the severity distribution of health states". It's defined as the average number of years that a person can expect to live in "full health" by taking into account years lived in less than full health due to disease and/or injury. These non-mortality measures, especially YLL, DALY and HALE, clearly need more detailed data for their calculation than death rates but offer useful summary information for monitoring populations. They are also clearly relevant to healthcare system performance despite having other influences.

4.6 Efficiency and Value

Interest in efficiency is growing with the awareness that, whereas we often think of poor quality as not giving patients the evidence-based treatments they need, limited resources are being squandered and patients harmed by the overuse of some tests and procedures. "Procedures of Limited Clinical Value" is a term NHS managers have applied to a range of elective surgical procedures that they no longer wish to fund. The list began comprising treatments considered to be complimentary or alternative, aesthetic, or without the National Institute for Health and Care Excellence (NICE) guidance of cost-effectiveness but was controversially expanded by many commissioners under pressure to make savings. In general, "unnecessary" treatment has been estimated to account for around 30% of U.S. Medicare costs (Fisher et al.

2003) in a much-cited study that compared spending with patient satisfaction and outcomes by region. In primary care, there has been concern over variations in (particularly high levels of) referral to secondary care (O'Donnell 2000) and in prescribing (Bourn 2007), particularly for antibiotics for viral respiratory disease (Dosh et al. 2000).

Metrics around efficiency and its cousin productivity are challenging to construct as they try to offer a comprehensive framework that links the inputs to the outputs, the resources used to the measures of effectiveness described earlier in the chapter. It's hard to do at a national level and just gets harder as one moves down from the macro to the meso-level, that is, to providers, departments and individual staff (Smith et al. 2009). As there was no consensus on how best to link inputs to outputs, the National Quality Forum commissioned a white paper to categorise the challenges of, and various methodologies for, doing this (Ryan and Tompkins 2014). It reflects the move towards "Value-Based Purchasing" to reward high-quality, cost-effective care.

The first task is to establish some definitions. We have already discussed what we mean by quality of care. The *costs* of care are defined by the NQF as "total healthcare spending, including total resource use and unit price(s), by payer or consumer, for a healthcare service or group of healthcare services associated with a specified patient population, time period, and [healthcare] unit". Healthcare spending is often expressed per capita population or as a proportion of the total GDP for the country; it's useful also to track affordability, a metric based on percentage of household spending on healthcare. This definition of cost is common but short-changes the patient, as it excludes things that are harder to get data on such as travel costs for treatment and opportunity costs. The latter cover the time and money that we spend on things that don't benefit us much which we could have spent on things that do benefit us. There's an opportunity cost if we don't take up preventive services like screening and then get the disease later on. There's also an opportunity cost to patients when they go for ineffective or unnecessary treatment or spend time navigating the medicolegal system. The NQF definition of cost also doesn't capture administrative costs for the provider or lost productivity because the patient can't work. Even the details of what is included can be complicated. For instance, costs under the NQF definition could be defined either as what the providers charge for their services (the bill) or as "allowed charges", which are negotiated between insurers and some providers. To get round the problem of price variation across the country, healthcare costs can also substitute average or "standardised prices". More broadly, costs arise from four areas: the health sector, patients and their families, non-health sectors (e.g., social care) and productivity changes (Drummond et al. 2005). Although many items in these categories have market prices, converting someone's time into money is harder, especially if that someone is a carer, a volunteer or the patient who has lost some of their leisure time negotiating the healthcare and perhaps also the legal system. An alternative to trying to put monetary values to these hours is just to add them up and present them in the results.

Efficiency of care measures the cost associated with a specific level of quality: this is more specifically known as technical efficiency. A unit is "technically efficient" if it produces the maximum amount of output from a given amount of input. A second perspective on technical efficiency is that such a unit needs the minimum amount of input to achieve a given level of output. Efficiency can also be assessed in terms of scale, price and allocative efficiency, but we will not consider those here.

Value is related to but not the same as efficiency. It measures a given stakeholder's preference-weighted assessment of a particular combination of quality and cost. As usual, the stakeholder could be an individual patient, a consumer organisation, a payer, a provider, the government or society in general. To monitor value, we need to monitor both cost and outcomes over time. Both are measured around the patient over a patient's entire care cycle. The question of what is value is discussed more fully by Porter (2010).

Another recurring theme in this book is that the choice of analysis should reflect the goals of the analysis. Quality and cost can be combined in different ways depending on the use. Common uses include internal efficiency profiling and improvement at a given healthcare unit, public reporting, pay-for-performance and health network design. For performance monitoring rather than for consumer choice, the intended user of the information is the healthcare unit. Focusing on "waste" (defined by the Institute of Medicine as including "waste of equipment, supplies, ideas, and energy") or "appropriateness" would therefore be relevant. There is no point in a unit offering a treatment at low cost without complications if it's of no benefit to the patient. Determining whether a given intervention on a particular patient was appropriate and not wasteful is notoriously fraught, and available data are often of insufficient detail to allow us to judge. Estimation of an appropriate level of, for example, use of imaging or specialist referral can instead be made by comparing the unit with a relevant benchmark. For instance, international comparisons have found that the United States does not deliver great quality of care compared with other rich nations, given its level of spending.

A large number of measures of efficiency have been proposed and were compared in the systematic review by Hussey et al. (2009). Almost all of them, however, turned out to be cost of care measures as they did not explicitly include quality. They are therefore of most use if quality doesn't vary much by unit. In their white paper (Ryan and Tompkins 2014), the NQF found seven models for combining cost and quality (Table 4.2).

They reviewed 25 programs, which used five of these methods to profile individual physicians, hospitals, health plans or whole systems; the "regression model" and "cost-effectiveness model" were proposed by Timbie and Normand (2008). Their report also uses real U.S. data to work through the methods in illustrative examples, which is useful if you want a deeper understanding. The authors conclude that there's no consensus on which approach to take. The commonly used methods have some limitations in common. They

TABLE 4.2

Ways of Combining Quality and Cost Measures in a 2014 National Quality Forum White Paper

Name of Model	Key Features	Comments
Conditional model/ net-incentive payment model	Four steps: (1) Assess quality; (2) assess costs; (3) classify performance on either or both domains, e.g., as low, average or high; (4) combine these classifications into one for efficiency	Private payers like this, e.g., to spot high-quality, low-cost providers
Unconditional model	First two steps are followed, but then quality and cost are scored independently and combined	Used in Medicare's Hospital Value-Based Purchasing program
Quality hurdle model	First three steps of the conditional model; then providers have to meet a minimum quality level – the hurdle – before their efficiency is assessed	Cost hurdle model is similar but the last step is the other way around
Regression model	Conceptually similar to the conditional model	Regression analysis accounts for the within-provider Correlation between quality and cost outcomes
Cost-effectiveness model	Assigns a monetary value to patient benefits from the specified quality domain	Can make a big difference to apparent performance
Data envelopment analysis/Stochastic frontier analysis model	The efficient frontier is modelled; providers' efficiency is then based on their distance from the efficient frontier	Can use continuous measures of cost and quality, so avoids arbitrary classifications of other methods. Results are sensitive to model specifications. Used a lot in research. See Section 4.6.1
Side-by-side model	Follows the first two steps of the conditional model and then displays the results for cost and quality separately.	Lets stakeholders place their own values on cost and quality

Source: Summarised from Ryan, A.M. and Tompkins, C.P., *Efficiency and Value in Healthcare: Linking Cost and Quality Measures*, National Quality Forum, Washington, DC, 2014.

generally use multiple measures of quality and so need to weight them and combine them into one composite. As we see in Chapter 8 on composites, different weighting possibilities exist and each can give a different answer. Second, the relationship between measured quality and patient health needs to be well defined but might not be. The implicit assumption is that if we can improve the quality while holding costs steady then we will improve efficiency. However, if quality is measured by a set of process measures that each impacts only a little on patient health, then large improvements in quality might not be seen in efficiency.

4.6.1 Data Envelopment Analysis and Stochastic Frontier Analysis (*)

The review by Hussey et al. (2009) found that the academic literature makes much use of data envelopment analysis (DEA) and stochastic frontier analysis (SFA) and that these are the two most common methods for assessing the efficiency of hospitals. They are therefore worth at least outlining.

Both bring together inputs (costs, numbers of staff, beds and service complexity, etc.) and one or more outputs (health service utilisation or quality of care). Both involve the concept of the *efficient frontier*, which comes from modern portfolio theory in the 1950s. When investing in companies, the efficient frontier is a set of portfolios that offers the highest expected return for a defined level of risk or the lowest risk for a given level of expected return. With healthcare units, it's the set of units that have the best outputs (quality) for their inputs (costs), so the lowest costs for a given level of quality or the highest quality for a given set of costs. With DEA, which can be called non-statistical or non-parametric, the frontier is jagged and ignores all sampling error. With SFA, an econometric method that uses regression, the frontier is smooth and accounts for sampling error, though this advantage is offset by the need for its underlying distribution assumptions to be met.

With the concept of the frontier comes that of relative efficiency: efficiency for a given unit is not measured in isolation but only in comparison with that of other units. DEA defines relative efficiency as the ratio of the weighted sum of outputs to the weighted sum of inputs. One obvious question is how to choose the weights. As units can legitimately value some inputs and outputs differently from other units, DEA lets each unit adopt a set of weights to show it in the most favourable light in comparison with the other units. Solving this is a mathematical linear programming problem. This method for choosing weights stops such arguments that are common with composite measures (Chapter 8), but a concern with DEA is that a unit can look good because of the choice of weights rather than being actually efficient. A minimum and maximum weight can be specified to limit this problem. Another reason why DEA is popular is that no distribution needs to be assumed for the frontier. It also handles multiple inputs and outputs. Accommodating several outputs is a useful feature, given that quality is multidimensional, though having more output variables than units is not unusual in DEA studies. Being non-parametric, it lacks the usual diagnostic tests for validity that parametric methods have, but options include bootstrapping, trying different combinations of input and output variables, assessing consistency of results over time and comparing the results with those from SFA when the same variables are used. DEA can be run in two modes: input-oriented and output-oriented. An input-oriented model holds the current level of output constant and minimises inputs, whereas an output-oriented model maximises output keeping the amount of inputs constant.

SFA defines relative efficiency differently. It uses regression to smoothly model the frontier, and then each unit's efficiency is estimated from the model's residuals, that is, their departure from this theoretical best practice, with statistical noise separated out. In this way, SFA assumes that differences between the actual performance of the unit and the optimal performance are not just due to the unit's inefficiencies but also to random factors that it can't control. This is a useful feature. Being a parametric method, though, means that it needs assumptions to be made about the functional form of the relation between input and output (known in economics as the production function) and also about the distribution of the error terms. However, a review of the use of SFA for U.S. hospitals found it to be fairly robust to technical choices such as the structure of costs and distribution of the error term (Rosco and Mutter 2007).

The choice of method should depend on the type of units of interest, the perspective taken and the quality of the available data (Hollingsworth 2003). Another rule of thumb is to use SFA when there's only one output and DEA when there is more than one (Parkin and Hollingsworth 1997). Both DEA and SFA have their pros and cons, though for both methods the selection of input and output variables is often the most influential on the results. This is of course dependent on data availability. Varabyova and Shreyogg (2013) outline the implementation steps for each method for cross-sectional and longitudinal analyses using OECD data. Their analysis of country-level data showed that, when the input and output variables are the same, the countries' efficiency ratings derived from the two methods usually correlated quite strongly. The results were more sensitive to the choice of variables in the models, particularly for DEA. For technical details of DEA and SFA, see, for instance, Jacobs et al. (2006).

The ideal output for either method would be some measure of the health gain of the unit's patients. Often, this isn't available, so for hospitals some measure of activity is typically used instead, such as discharges or inpatient bed days. Discharges are probably the better of the two as delays at any of various points during the patient's stay represent inefficiency so that more would not mean better. AMI mortality is a popular choice of outcome measure in efficiency studies of hospitals and has been used for hospital benchmarking within and between OECD countries.

The review by Hussey et al. (2009) found that the grey literature on healthcare efficiency is different from the academic in several ways. The methods used have care episode groups, for example, episode treatment groups as outputs, which are usually ratios, such as observed-to-expected ratios of costs per episode of care, adjusting for patient risk. The details are proprietary as they belong to a vendor. The authors noted limited documentation of assessment of the reliability and validity of measures and concluded that the state of the science regarding measuring efficiency is some way behind that of measuring other aspects of quality.

4.7 Features of an Ideal Indicator

There is no such thing in practice, but it's useful to begin by drawing up a wish list of features of an indicator. Mainz (2003a) gave these key characteristics: "(i) indicator is based on agreed definitions and described exhaustively and exclusively; (ii) indicator is highly or optimally specific and sensitive, that is, it detects few false positives and false negatives; (iii) indicator is valid and reliable; (iv) indicator discriminates well; (v) indicator relates to clearly identifiable events for the user (e.g., if meant for clinical providers, it is relevant to clinical practice); (vi) indicator permits useful comparisons; and (vii) indicator is evidence based". Pringle et al. (2002) produced a longer set of attributes, given in Table 4.3.

Any measure will meet the ideal attributes in Table 4.3 to a greater or lesser extent and require value judgments in deciding. It's easier to describe the psychometric or technical performance of an indicator, such as the amount of random variation in a rate expected under the binomial distribution, than to assess the influence of contextual effects on an individual's performance.

4.8 Steps in Construction and Common Issues in Definition

Our reasons for monitoring quality are influenced by interrelated factors such as government policy and/or targets and perceived or known need for improvement. These in turn help us decide on the clinical area that our indicators should assess. It makes sense to focus on healthcare problems or diseases with high volumes, morbidity, mortality and associated treatment costs. If the monitoring is for improvement, then it also makes sense to choose areas where healthcare is known to influence outcomes more strongly and where variations in quality are known or at least thought to exist. Such an evidence base will be much sparser for rare conditions. A systematic review of methods for choosing indicators found that the two most common selection criteria were the public health importance and the gap between actual and potential performance (Kötter et al. 2012). We'll consider this issue in more detail in Section 4.10.

Structure and process measures are typically derived from clinical guidelines that bring together the evidence supporting the performing of certain processes in caring for a particular type of patient. Prescribing aspirin on discharge for those hospitalised with acute myocardial infarction is a process measure based on strong evidence that aspirin can prevent future cardiovascular events. The underlying evidence will typically

TABLE 4.3

Attributes of Ideal Quality Indicators

Attribute	Description	Comments
Validity	Meeting the standard is seen as better quality.	Psychometrics recognises various types of validity, e.g., face and content.
Communicable	Relevance of measure can be explained to target audience.	Other authors use the term "meaningful".
Effective	Indicator measures what it purports to measure and can be used for its desired purpose.	It should also be free of perverse incentives and unintended negative consequences.
Reliable	Underlying data are complete, accurate and consistent, and indicator is reproducible.	Measures with lots of statistical or other noise will score poorly here.
Objective	Data are independent of subjective judgment.	Not important if we want patients' views of their care experience.
Available	Data are available quickly/routinely with minimal cost or effort.	
Context-free	Indicator is context-free or important context effects should be adjusted for.	
Attributable	Performance on an indicator can be attributed to the relevant individual or team/unit.	See Chapter 3.
Interpretable	How well the measure reflects quality of individuals or units must be explicit; measure should be used appropriately in its presentation and interpretation.	Rates of GP referral to hospitals don't just depend on GP's abilities.
Comparable	Indicator should be comparable against a gold standard, e.g., NICE guidelines.	Also see Chapter 9 on defining outliers: "good" or "poor" compared with what.
Remediable	Poor performance on an indicator can be remedied using recognised and feasible methods.	Some effects due to socioeconomic deprivation and lifestyle are likely to be beyond the powers of the healthcare system to fix.
Repeatable	Indicator should be sensitive to improvement over time.	Measures with lots of random or otherwise unpredictable variation will score poorly on this.

Source: Attributes and descriptions taken and adapted from Pringle, M. et al., *Br. Med. J.*, 325, 2002.

come from in vitro and in vivo studies and is likely to vary in quality (there are tools for assessing study quality). This evidence is typically then converted into clinical guidelines by a panel of experts for most of such patients in most circumstances. Guidelines are mostly developed for single conditions. The use of the recommendations in these guidelines in diagnosing and treating individual patients is then left to the discretion of the physician, who must take account of, among other things, the common and inconvenient fact that the patient in front of him or her has more than one condition, as we saw in Section 4.4. For more details on guideline development, see the series of *BMJ* articles (Shekelle et al. 1999) or the international review of handbooks (Turner et al. 2008). An introduction to the basic steps involved in developing indicators from guidelines is given by Mainz (2003a). The procedure is, however, more complicated than this, with various methods in use. Even among working groups specialising in guideline and indicator development, a variety of methodological approaches exists (Blozik et al. 2012). A review found that often one but sometimes several guidelines were used to derive indicators. Twenty studies detailed other sources, such as existing indicator databases, in addition to clinical guidelines (Kötter et al. 2012).

However, guidelines are about recommendations, but they do allow physician discretion. That is not what an indicator does. An indicator will imply that unless that every AMI patient is given aspirin on discharge, the doctor has made a mistake. This ignores contraindications, patient preferences and conflicts due to comorbidities and other patient factors. There is a clear need when guidelines are constructed to consider at the same time how a related indicator should be constructed that monitors performance in relation to them (Quality of Care and Outcomes Research in CVD and Stroke Working Groups 2000). For instance, carefully excluding patients with contraindications will reduce or eliminate the need for risk adjustment, a key advantage that processes hold over outcomes as discussed earlier and in Chapter 6. Such refining of the denominator will make the process measure more specific and less "noisy" but can be difficult to achieve with available data and can result in small sample sizes.

Outcomes are states of health or events that follow care and that may be affected by healthcare (Mainz 2003b). The ideal outcome measure would be solely determined by care processes and not by patient factors. No such entity exists. Even direct complications of care, such as intraoperative organ injury, are more likely in some patients than others. Chapter 6 is devoted to the adjustment for patient factors (casemix). An ideal outcome indicator would capture the effect of care processes on patients' health and well-being and be sensitive to variations in those processes. Mortality is commonly used, and we've already seen different versions of it for monitoring population health that acknowledge influences other than healthcare. For hospitals, however, we're mainly interested in the proportion of deaths that are deemed preventable with up-to-standard care at a given hospital. Although studies estimate this proportion to be around 5% for in-hospital deaths, the definition used is often very strict. A second problem is the finding that the method for determining preventability – casenote (chart) review – is expensive and of modest reliability

(Sari et al. 2007; Hutchinson et al. 2010), though it can be improved upon with modifications (Hutchinson et al. 2013). Estimating the number of "preventable deaths" in a given sample is fraught with difficulty (Hayward and Hofer 2001). Even with a higher estimate, however, a large chunk of the total will not be preventable no matter how good the patient's care is. The challenge for those aiming to construct an outcome measure is to maximise its sensitivity to the effects of healthcare, the flip side of what's done when constructing process measures. Beyond such technical considerations, outcomes should be considered important to users and often also to society at large.

4.9 Validation of Indicators

For structure and process measures, validation is generally done by establishing a relation with outcomes (Donabedian 1966; Palmer and Reilly 1979): good care yields good results. The inherent problem with this is that outcomes are also influenced by factors that have nothing to do with care. Observed correlations between process and outcome measures can therefore be modest or even non-existent. There are several possible reasons for this:

- The process measure does not in fact measure quality.
- The outcome measure is not relevant to this group of patients.
- Both measures are relevant to quality but they measure different dimensions of care.
- The outcome measure is relevant but insensitive to quality due to a real but simply weak impact of the process on the outcome.
- There is a real relation between the two, but the study was underpowered to find it, perhaps because the outcome was rare.

Distinguishing between these is hard. The chicken-and-egg nature of the "outcomes validate processes" chicken and the "processes lead to outcomes" egg receives little attention, whichever one you take to have the feathers. However, most authorities argue that outcomes are the most important focus. As just one example, the mission of the not-for-profit International Consortium for Health Outcomes Measurement (ICHOM) is to create condition-specific standard sets of outcome measures and drive their adoption and reporting worldwide. Providing that these have been chosen carefully, in consultation with all the key stakeholders as in the ICHOM approach, they should be the gold standard.

4.10 Some Strategies for Choosing among Candidates

Preferably, each candidate indicator has a research evidence base to support its candidacy rather than expert opinions or clinical experience alone. The evidence can be rated for quality using different scoring systems, for instance, A to D, where A involves at least one randomised controlled trial and D simply expert opinions (Eccles et al. 1998). The use of such a rating allows transparency.

A common but intensive approach to reduce a large list of candidates is to gather an expert panel. The procedure involves some modification of the Delphi method as formulated by the RAND Corporation over 50 years ago. This essentially means two or more rounds of the experts giving their ratings of each candidate indicator, with the least popular culled in each round until a consensus is reached; the RAND method specifies what constitutes consensus. The result from this should have good buy-in from fellow professionals, though that's not guaranteed from other stakeholders such as patients and commissioners. Many panels also therefore include patient and administrator representatives.

Sometimes, the intention is to compare performance with that in other countries. For example, the committee selecting indicators for the first annual report in 2015 by the National Healthcare Quality Reporting System in Ireland wanted to align theirs with those abroad, which often meant comparing their results with those published by the Organisation for Economic Cooperation and Development (OECD). This will inevitably reduce the shortlist. More generally, data availability and feasibility of measurement will have a big say. We should start by assessing whether existing data and existing indicators are fit for our purpose. If what we have is considered to be too narrow in focus or if pilot testing shows that some indicators don't vary much across units, new indicators will need to be developed. However, you can still choose to retain an indicator that's considered crucial for user acceptance of the monitoring system even if the between-unit variation is limited or other aspects of it are less than ideal. This point will be picked up again in Chapter 8 on composites, which also includes some discussion on indicator selection.

Another strategy to choose indicators involves estimating their relative merits in terms of population health gain. Health gain could be measured in quality-adjusted life years (QALYs), for example. Achieving optimal performance on indicator 1 might only add 10 QALYs, whereas achieving optimal performance on indicator 2 might add 100 QALYs. We would therefore be more interested in performance on indicator 2, especially if the necessary quality improvement projects cost the same amount to implement. Meltzer and Chung (2014) considered the implementation costs and the health gains in an analysis of 13 AHRQ National Healthcare Quality Report indicators. They used the concept "net health benefit". This refers to the gains in health from a given intervention compared with an alternative intervention, after subtracting any health improvements that have been lost because of the costs of the intervention. The gains in health can just be estimated from the difference in QALYs between the two interventions, with figures taken from published cost-effectiveness studies on the relevant population. The costs of the "lost" health improvements are converted into QALYs by assuming a dollar equivalent, typically in the range of $50,000 – $100,000 in the United States. In the example that they give, with a threshold of $100,000 per QALY, an intervention producing ten QALYs but costing $100,000 would produce a net health benefit of nine QALYs. Their framework is based on having only one alternative intervention, but they discuss how to extend that. Also needed for the calculation is a set of denominator populations, that is, how many people are eligible for each particular intervention. For some indicators such as screening or immunisation, this can be the whole population or big chunks of it, so the census can be used; the authors list their information sources. They found that perfect implementation of the top seven measures would yield 93% of total QALYs possible from their set of 13. The percent of adults with diagnosed diabetes with most recent blood pressure <140/80 mmHg was easily top. In contrast, six measures, such as prescribing for hospitalised heart failure patients with left ventricular systolic dysfunction, accounted for the small remainder. The bottom two related to acute care patients, so had relatively small denominator populations. Within this framework, the biggest potential net gains would be those conditions that affect large numbers of people, where treatment makes a big difference to each person's health, and where the gap between what's now achieved and what's in theory possible is great. The authors acknowledge that they lacked information on how much effect quality improvement efforts could have in reality – some things are far easier to improve than others – and on the costs of such efforts. Still, even if that is not available, their framework looks promising for process of care indicators if the other data can be obtained.

4.11 Time to Go: When to Withdraw Indicators

Indicators should be periodically reviewed. There are several reasons why we might want to modify or even withdraw them from use. Policy priorities and the clinical evidence base can change over time. If performance improves to the extent that it reaches and maintains a ceiling, then the focus can switch to other measures. An exception to this occurs in a pay-for-performance (P4P) scheme when the financial incentive pays for extra staffing that's needed to maintain the high performance level. Withdrawing the indicator and hence the extra money that comes with it will lead to lower achievement (Kontopantelis et al. 2014). If performance against the indicator is high, it does not necessarily mean that everything is well. P4P schemes and declared targets have the effect of distorting staff behaviour, which can include gaming and cream-skimming. Gaming comes in many forms, but the main element is manipulation of the situation or the data to make performance appear better than it actually is. One type of gaming is cream-skimming, which refers to choosing patients for some characteristic(s) other than their need for care; this typically means choosing patients who are less severely ill and therefore cost less to treat. Organisations and individuals naturally focus on what's measured, particularly what's publicly reported and/or incentivised, at the potential expense of the rest. Bevan and Hood (2006) give examples of gaming and four possibilities for what's really happening when some performance target is apparently met:

1. All is well in all areas, whether measured or not.
2. All is well where performance was measured but not where it wasn't.
3. Performance against targets seems to be fine, but actions haven't been in the spirit of the goals behind those targets (the unit has missed the point of the target).
4. Targets have been missed, but this has been concealed by ambiguity in the way data are reported or by outright fabrication.

An example of the third point relates to the target that most patients in the Emergency Department (ED) should wait no more than 4 hours (in England, this target is currently at 95% but peaked at 98% in earlier years). As the clock starts only when the patient books in at the ED's reception, sometimes patients had to wait in ambulances outside the department until staff were confident of meeting the target (CHI 2005). It's impossible for users of the information to distinguish between these four very different possibilities without further knowledge. Inspection and data auditing can combat such problems, and changing the indicator set so that they cover other aspects of care may help tackle gaming (Kontopantelis et al. 2014).

4.12 Conclusion

Whatever we decide to measure, as seen with ED waiting times and P4P schemes given earlier, it's vital to consider the potential positive and negative impact on human behaviour. The impact of any performance monitoring scheme must consider the unintended consequences not only around our chosen set but also around areas that we have decided not to, or have been unable to, measure. Chapter 12 will pick this point up in more detail.

5

Sources of Data

Aims of This Chapter

- To describe how to assess the quality of the data
- To give an overview of the main types of data source for performance monitoring, together with their strengths and limitations
- To flag up other important issues such as linkage and ethics

It can be said that the aim of performance monitoring is to turn data into information. As Daniel Keys Moran put it, "you can have data without information, but you cannot have information without data". This chapter concerns information's raw materials. It's impossible to list all the data sources, but they can be broadly categorised as administrative, clinical, incident reports, surveys and a variety of other types such as social media and population estimates. The suitability of each for the intended purpose will partly depend on their quality. As we'll discuss, this can be defined in different ways but ultimately involves a subjective judgment of whether the source is "fit for purpose". Are the data "good enough"? As our discussion of data sources will use various terms associated with data quality, we'll begin by defining those terms to help with the summary of the data sources that then follows. We finish by considering a few other important issues such as ethics and costs.

5.1 How to Assess Data Quality

First, we need to define this, which can be done in different ways. The most common term is surely "accuracy". In this chapter, we'll happily use it as a shorthand or where the specific component of data quality doesn't matter, but often the term isn't detailed enough. Three old measures are still in use and are still helpful: coverage, completeness and accuracy itself. These can be used in different ways by different people, but here are our definitions. Suppose that we want to construct a register of residents of a town and determine each resident's smoking status. We send out a paper postal questionnaire and ask people to fill it in and return it to us, after which we'll input the information into our tailor-made electronic database. How good is what's on that database? Coverage describes our database's ability to list every resident in the town, irrespective of what we've put down for their smoking status. In this example, if we're using the electoral roll to get a list of names and addresses of the residents, then coverage will be 100% only if every resident is registered to vote. If instead we've no such list of names and addresses and we get someone to hand-deliver the questionnaires to each address in the town, or if we hand them out in public places such as clinics, places of worship and shopping centres, then we're relying on people to return them. Completeness is our ability to put something in the "do you smoke?" box on the form, even if it's wrong. Accuracy is our ability to get that something down correctly so that it reflects whether each person really does or does not smoke in reality. If we've used the electoral roll but many people aren't registered to vote, or if lots of people don't return the hand-delivered questionnaires, then coverage will be low. If lots of them return it but don't fill all of it in, then completeness will suffer. If we make lots of data entry errors – or many of the residents lie about their smoking habits – then accuracy will be poor. Using this definition, a database will therefore be highly accurate if all the entries reflect reality for the set of residents who returned the questionnaire. Of course, if only a handful return it, the database is still useless, however correct their answers are.

As we'll see when we discuss the literature on the reliability of administrative data, other terms taken from epidemiology and screening are also useful. If we want to use a particular source to estimate the prevalence of a disease, for instance, what proportion of inpatients have diabetes, then it's helpful to know how reliable this estimate is. We can use administrative data on hospital admissions and note the proportion of patients with diabetes recorded in any of the diagnosis fields, but for various reasons that we'll come on to later this may likely be an underestimate. The gold standard source would be the patient notes or charts, which may be electronic (for illustrative purposes here, let's ignore any mis- and under-diagnosis of diabetes). If the chart says diabetes but the electronic record says not, this would count as a false negative for the electronic record. The electronic record's sensitivity – its ability to pick up cases of diabetes – will be less than 100%. This is common. If, however, the chart says no diabetes but the electronic record does say diabetes, this will be a false positive for the electronic record. Its specificity – its ability to pick up people who are not cases – will be less than 100%. A third measure is also useful: of those patients who the electronic record has labelled diabetic, what proportion actually have diabetes, at least according to their chart? This is termed the positive predictive value (PPV). In media coverage of screening tests in particular, the term accuracy is banded about to refer to how good a screening test is, but without further specification it's pretty meaningless. The most useful definition will depend on the context.

Another angle comes from psychometrics. "Reliable" is another characteristic that we'd like our data to have, but what does it mean? There are different components to it. Test–retest reliability describes to what extent a measurement made on the same element at different times gives the same answer. If we take Mrs Jones's blood pressure at the start of her consultation, find it to be 155/95 and label her hypertensive enough to warrant medication, how likely are we to want to prescribe her antihypertensive medication when we take her BP at the end of the consultation? Another common component, interrater reliability, describes to what extent different people or instruments (raters) give the same measurement on the same element. If you take Mrs Jones's BP instead of me, would you also get 155/95 for the first reading? For many elements in medicine, and this includes patient notes (see end of Section 6.4.3), our gold standard is silver at best. We therefore have to take some average of multiple measurements, perhaps using multiple raters, and hope the reliability is high. If it's known to be low, such that the instrument has appreciable measurement error or that clinicians differ greatly in their opinions, then more readings/raters are needed before we can trust the average.

Data quality problems can have a big impact on an individual patient's predicted risk of some outcome if we build a statistical model that includes a number of unreliable and inaccurately recorded data items. This is one reason why risk-prediction models developed using a single institution's data don't perform so well when applied to those from another institution. Errors can also affect the performance measures themselves for an individual patient. However, when data are aggregated by unit, there is a tendency for many errors to cancel each other out to some extent. Of course, if unit A has consistently defined infection using different criteria from unit B, then no amount of aggregation will magic the problem away. But, in general, particularly for risk-adjustment items, data quality does not need to be as high when comparing units as when predicting a given patient's outcome at the bedside for the purpose of clinical management. Ultimately, we have to decide whether the data we're proposing to use are fit for our purpose. This will necessarily involve some of the technical measures just discussed, but it will also consider other factors such as availability and timeliness together with a subjective assessment of what level of inaccuracy, in all its forms, we and the users of our monitoring system are happy to tolerate.

5.2 Administrative Data

Administrative data are routinely recorded without any specific research question in mind. These might include data collected for legal, civil registration, financial or insurance purposes. Many Westernised countries have extensive electronic systems with standardised collection procedures for regional or national aggregation. In many countries, the registration of births, marriages and deaths dates back hundreds of years in paper format. The World Health Organization (WHO) describes the registering of all births and deaths and the recording of causes of death as "one of the most valuable assets a country can have" (WHO 2014). Our favourite nurse-epidemiologist, Florence Nightingale, recognised the value of standardisation and, with the aid of physicians such as William Farr, developed an ambitious Model Hospital Statistical Form. This aimed to capture the number of patients in hospital at the beginning and end of a year, the number of patients admitted during the year split by outcome (recovered, discharged as incurable, dismissed at their request, died) and even the mean length of stay. Alas, this did not achieve widespread use, partly thwarted by Farr's disease classification that put him at odds with many pathologists (see the essay by Cohen at http://www.unc.edu/~nielsen/soci708/cdocs/cohen.htm). Nonetheless, Farr and Marc d'Espine were asked in 1853 by the International Statistical Congress to create a cause of death classification that would later form the basis of the ICD system, which we discuss later. Nightingale also saw the value in comparing hospital performance on the basis of such records and to see "whether [subscribers'] money was not doing mischief rather than good". She thereby invented billing data.

Administrative data can go by other names, depending on their origin. The term routine can be considered synonymous, whereas "billing" or "claims" are specific to health insurers or to systems that use some kind of fee for service so that the hospital or physician has to bill or make a claim to some funder or purchaser. We will generally use the term administrative or routine as we don't want to limit ourselves to such hospital- or office-based records. Performance monitoring will cover other healthcare sectors and make use particularly of information on population counts, births and deaths and perhaps also crime, education, welfare and environmental sources, all of which can often be considered routine.

As their name suggests, administrative data are collected regularly for some statutory or financial purpose, but to fulfil their primary role they need to be detailed. It's the combination of detail, standardised collection and wide coverage – often regional or national – that makes them so attractive for researchers and monitors. A birth registration is a legal process, but the record will often capture elements of wider usefulness than just the name and address, such as birth weight, gestational age and birth outcome. The electronic systems, formats and procedures for accessing and analysing the data are relatively cheap and often well established, which means that there are a lot of people with experience in handling them. With this experience comes an awareness of their main limitations, such as lack of clinically useful detail (particularly around patient physiology), variable data quality and in many cases a time lag between the occurrence of the care event and the user getting their hands on the records for that event. At worst, this time lag can be measured in years. Missing or simply wrong data can be common, especially if the time available for data entry is limited and if there is no external assessment of the quality or penalty for poor standards. If certain data items are linked to payment, as is the case in many countries for diagnosis and procedure fields for hospital admissions, then one can expect the standard to be fairly high. If the item is recorded accurately, however, it can still be frustrating to learn that it isn't defined the way you need it. Always keep in mind that these data were not designed specifically for your monitoring scheme. The strength of analysis of administrative data is identifying patterns and formulating hypotheses rather than giving definitive answers to variations in performance. These features are what make them so popular with performance monitors around the world.

5.2.1 Coding Systems for Administrative Data

Routine data rarely use free text, except for personal details for births, marriages and deaths, for instance, which are always removed before the database is made available to third parties. For ease of analysis, coding systems are employed for such information as diagnoses, operations and procedures, medications, ethnicity and type of residence. We will only consider here the first two on the list. The standard but not the only way that the world classifies diseases and other health problems is WHO's International Statistical Classification of Diseases and Related Health Problems, or International Classification of Diseases (ICD) for short. The first version has a rather complicated history (see http://www.who.int/classifications/icd/en/HistoryOfICD.pdf), but in brief, delegates from 26 countries attended a conference in 1900 and adopted a detailed classification of causes of death consisting of 179 groups and an abridged classification of 35 groups, with recommendations that the system be revised every 10 years or so. The tenth revision came into use in the WHO Member States from 1994, with ICD11 due out in 2017; the United States was very late to make the change to the tenth revision. The thousands of alphanumeric codes are organised mostly by anatomy or body system. For ICD10, for instance, two chapters of infections are followed by neoplasms, and blood and endocrine diseases.

For some countries, including the United States, Canada, South Korea, Germany and Australia, standard ICD is not considered sufficiently detailed, and they have added further subcodes into a "modification" version such as ICD-CM ("clinical modification") in the United States and ICD-AM ("Australian modification"). ICD-CM is much larger than standard ICD and is updated more frequently (each October). For example, the National Institutes of Health Stroke Scale was added as an extra digit to acute ischaemic stroke codes in October 2014 to capture stroke severity.

Each ICD revision comprises thousands of codes, which often need to be grouped to make meaningful entities. A disease will typically be divided among several codes. Asthma with its various subtypes in standard ICD10 is

J45, with status asthmaticus (acute severe asthma) J46; ICD10-CM has several codes for the latter appended to J45 instead. Multiorgan conditions such as diabetes are served by a larger number of codes. Users of administrative databases need to choose the definition of their index condition, for example, E10-E14 for diabetes and I60-I64 for stroke in ICD10 are common but not universal. The WHO themselves provide a number of classifications among their tabulations of morbidity and mortality, but users typically need more granularity. A widely used multipurpose grouping of ICD9 and ICD10 codes was developed by the Agency for Healthcare Research and Quality (AHRQ) in the United States. This Clinical Classifications Software condenses the 14,000+ ICD10 codes into 260 clinically meaningful conditions for health services research. It's a good starting point that can be adapted for your specific situation if needed. For ICD-9-CM, the CMS contracted physicians and statisticians to put the 15,000+ codes into 189 Hierarchical Condition Categories (Pope et al. 2004), which have been the basis for reimbursement for Medicare Advantage plans (Medicare Part C) since 2004.

For procedures and operations, the picture is more complicated. As part of ICD-9-CM, the U.S. National Center for Health Statistics created Volume 3 for procedures, which was replaced by the ICD-10 Procedure Coding System (ICD-10-PCS). ICD-10-AM includes an Australian classification of procedures based on the Australian Medicare Benefits (fee) Schedule. Other countries have also produced their own classifications for procedures, making international comparisons difficult. A further problem is that, unlike for diagnoses where we have groupings by the WHO and AHRQ, there is no standard way to group codes into meaningful procedure definitions. This is therefore done on an ad hoc basis and can generate a lot of debate among surgeons. See Chapter 10 for a worked example of doing this across different countries.

As routine data are often used for reimbursement, a method of classifying admissions but also other healthcare contacts into entities of equal cost is needed. The most common is the Diagnosis-Related Group (DRG) system (the United Kingdom uses the similar Healthcare Resource Group system [HRG]). All healthcare contacts for a given DRG are assumed to cost the same amount of money (they don't, but they all imply the same charges to the provider). Hospital resource use is dominated by what operations the patient had or what their primary diagnosis was. Other factors that the system typically accounts for include age, comorbidities, complications and length of stay. There is a clear financial incentive to fully record these factors, which generally leads to better data quality, particularly at those units with the resources to focus on this issue. There is the potential problem of over-recording in some cases: this has long been termed "DRG creep" (Simborg 1981). The temptation to bend the coding guidelines exists when the data are used not only for reimbursement but also for public reporting. If units know that the risk-adjustment model makes use of the presence of certain elements in the data, such as comorbidities or the use of palliative care services (Chong et al. 2012), there is a possibility for upcoding and/or gaming. We'll discuss the relative accuracy of administrative and clinical data, particularly regarding comorbidities, after the next section.

Even in the absence of pressures on recording such as public reporting of performance, there has long been concern in some quarters over the quality of some important fields in administrative databases. Studies in the 1990s and earlier are much less relevant to us today, as widespread problems with IT limitations and the fact that the data being collected simply were not used or scrutinised that much then led to a bad reputation that was often merited. Our research group's systematic review found significant improvements this century since those earlier studies: since 2002, the accuracy of the primary diagnosis was found to be 96% and for the primary procedure 84% (Burns et al. 2012). Accuracy is enhanced if three- rather than four-digit codes can be used, as many errors occur at the fourth digit. There are two separate issues regarding the recording of diagnosis and procedure codes. One is whether the information has been correctly entered according to the established coding guidelines. That is what "accuracy" should refer to in order to be strictly fair, though comparisons are usually made using chart review. Physicians and medical records staff don't always speak the same language when it comes to coding, which is nicely illustrated in the vignettes by Ballentine (2009). A second issue is whether what's been entered, even if correct according to coding guidelines, is useful to us the user and in particular whether it's useful for performance monitoring. When judging these two issues, we should also bear in mind a third one: whether the patient has been diagnosed correctly by the physician (see section on clinical data), which is often ignored in studies of data quality when chart review is used as the gold standard. One study that addressed this, used 45 trained standardised patients who presented with four common outpatient conditions to a large, U.S. private medical centre (Peabody et al. 2004). The correct primary diagnosis was recorded for 57% of visits; 13% of errors were caused by physician diagnostic error, 8% by missing forms, and 22% by incorrectly entered data. Secondary diagnoses fared worse. Among these overall figures, there was considerable variation by diagnosis and hospital site but not by level of physician training. Entry errors were much higher when the clerks entered the data than when the physicians did. Surprisingly, the hospital site without financial incentives for accuracy, such as reimbursement by fee for service, had worse results than those sites with such an incentive. The authors attributed this to the fact that the paper forms were not entered electronically at the time care was given. Local factors can always trump national ones. There are some limitations to the scope of the study in terms of hospitals and conditions covered, and the authors emphasise that their findings should not be generalised uncritically to the accuracy of all administrative data. The level of missing forms was high here, and several systems that are already in place elsewhere were suggested that would greatly reduce the non-physician errors. In the United Kingdom, for example, medical paper casenotes (patient charts) are converted into electronic records that include ICD10 and other coding systems by trained clinical coders. As the Professional Association of Clinical Coders website points out, the role is that of a "translator of clinical terminology as opposed to a clinical practitioner". The National Clinical Coding Qualification is the nationally recognised qualification for NHS coders. It's awarded after the passing of a theory and a practical exam, which includes questions on anatomy and physiology as well as various coding systems.

5.2.2 Use of Administrative Databases to Flag Patient Safety Events

Given their advantages, particularly of low cost and high population coverage, administrative data are a potentially appealing source of information on patient safety. Most of the early work on patient safety in the 1990s used casenotes (chart reviews). An early review of what administrative data could offer argued that they could screen for potential patient safety problems that merit further investigation and assess the adverse impacts of these incidents (Zhan and Miller 2003). Regarding performance monitoring, they wrote, "To some extent, [they can] provide benchmarks for tracking progress in patient safety efforts at local, state, or national levels". The most extensively used tools in this space are the Agency for Healthcare Research and Quality (AHRQ) patient safety indicators (PSIs), developed in the United States in 2002 and subsequently adapted for use in a number of countries such as England (Bottle and Aylin 2009), various countries in mainland Europe by the OECD, Australia (McConchie et al. 2009), and as a WHO version (Quan et al. 2005). Development included literature reviews of candidate indicators, clinical and coding expert review, empirical analysis and the development and documentation of software tools. Among the original set of 20 indicators were measures around infections, pressure ulcers and obstetric tears, but they are restricted to events that have corresponding ICD9 or ICD10 codes. The software produces crude and risk-adjusted rates.

A number of mostly single-institution studies have since assessed the sensitivity, specificity and positive predictive value (PPV) of these PSIs. The results have been mixed. The PPV has been good for PSI15 accidental puncture or laceration (Utter et al. 2009), PSI11 post-operative respiratory failure (Utter et al. 2010), PSI7 selected infections due to medical care and PSIs 18-20 obstetric trauma (Bottle and Aylin 2008b), moderate for PSI2 death in low-risk DRGs, PSI5 foreign body left during procedure and PSI13 post-operative sepsis (Bottle and Aylin 2008b) but moderate or poor for PSI3 decubitus ulcers (Polancich et al. 2006; Bottle and Aylin 2008b). The authors noted that the PPV could be increased by recording extra minor cases and that the cases flagged as PSIs were not necessarily preventable. PSI and their false positive rates could also be inflated by lack of present on admission flags so that comorbidities were mistaken for complications (Bahl et al. 2008).

A few studies have been able to compare the PSIs when derived from different databases. Koch et al. (2012) looked at post-operative complications in three sources: the AHRQ administrative database, a national clinical registry (National Surgical Quality Improvement Program [NSQIP]) and an institutional clinical registry (Cardiovascular Information Registry). Agreement (concordance) was low except for deep vein thrombosis, with differences in definition given as the main but not sole explanation. The agreement between the two clinical databases was only slightly higher than that between either clinical database and the administrative source. This questions the use of NSQIP as the gold standard in a similar study that also found low agreement (Cima et al. 2011).

The developers and many users of PSIs express caution about the public reporting of unit-level results, recommending instead that PSIs be used internally for quality improvement. Nonetheless, public reporting is already widespread in the United States. With performance transparency being increasingly seen as a vital tool in quality improvement, qualms over limitations and variations in coding are likely to take second priority.

5.3 Clinical Registry Data

An alternative type of database is purpose-built, generally by professional organisations such as the UK Royal Colleges or specialist societies, by U.S. states, by hospitals or by private groups. A number were started by enthusiastic individuals. These databases are specific to the specialty and so cover a narrow range of diseases and procedures. Types include national audits, birth cohorts, disease registries and procedure registries. Some areas of medicine are covered more than others, with cancer and surgery more than mental health, for example. Freed from statutory or other administrative requirements, these databases can contain a wealth of extra clinical detail that in principle can lead to more useful measures of quality and safety and better casemix adjustment. They have also been designed for the explicit purpose of monitoring and improving quality of care. Cases recorded by registries are likely to fulfil international diagnostic criteria, limiting false positives. Face validity and buy-in from users, particularly physician users, is likely to be high as a result. Data entry is usually done by nurses or occasionally by the clinicians themselves. Staff collect the data using a standardised paper form from sources such as patient notes and lab results and type the information in to create the electronic record. This can incur transcription errors and is expensive. For example, the National Surgical Quality Improvement Program is the leading outcome-based program to measure and improve the quality of surgical care in the U.S. private sector. The current cost of implementing NSQIP in a private sector hospital was put at approximately $135,000 per year (Kammermeister 2009), most of which is on the salary and training for the nurses who collect the data by abstracting patient charts. In the United Kingdom, it has been estimated that clinical registry records cost between 10 and 60 times more than getting data through hospital administrative systems (Raftery et al. 2005). Much better is to abstract the information from already existing electronic sources that are generated in the course of normal clinical care, though rates of adoption of electronic health records (EHRs) remain low even in the United States (Jha et al. 2010). As with administrative data, one can only abstract or use what's been documented, whether (legibly) hand-written or inputted. EHRs will retain values of variables forever unless someone updates them. Copying and pasting information from another source or appointment also serves to perpetuate errors in documentation.

Regardless of whether the format is paper or digital, the estimated level of performance and the success of any risk adjustment involved depend on the accuracy of the recorded information. Just because it's being used for patient management doesn't mean that it's infallible. Interrater reliability is key here – the medical literature is full of evidence of disagreement between healthcare staff over what test results mean or whether a physical finding is present. Staff might not even agree with themselves when assessing the same evidence at different time points as we saw earlier in our hypothetical example of measuring blood pressure. The level of disagreement can also depend on the level of the staff's experience. The Journal of the American Medical Association (JAMA) has a series of articles on the quality of the evidence around various clinical conditions and is available at http://jamaevidence.mhmedical.com/book.aspx?bookId=845. This relates to information collection by staff, but there is also a lot taken directly from the patient by asking them about their symptoms and social circumstances. This is usually done face to face, but some places get patients to enter this into a terminal or online. Patients also have growing access to their own EHRs. Such self-reported data are subject to influences of language, education, socioeconomic status and culture. A lot of this information from both staff and patients is potentially unreliable, but that doesn't necessarily invalidate its use as either a performance measure or as a risk-adjustment variable. We have to consider whether there are likely to be systematic differences across units that bias our results. Good evidence on this is generally lacking. A U.S. study found that when a complication occurred, staff documented what they did and what they found more thoroughly than for patients in whom no complicated occurred (Iezzoni et al. 1999). The same researchers also found that levels of information recorded varied by teaching hospital status. Levels also related to how well the physician knew the patient from previous encounters and how much they needed to inform their colleagues using detailed notes. The frequency of documentation can also vary by hospital size, as bigger hospitals with more staff write more times during each day than smaller hospitals with fewer staff. Having more recordings leads to discrepancies, partly because some symptoms like wheeze vary in intensity over time. All these issues happen with paper records and EHRs alike.

Submission to clinical databases can be voluntary, which can of course affect levels of coverage and completeness (see Section 5.4). There can be a charge to participate in the registry, and the fee structure can put off some types of units more than others. A 2004 review of 105 multicentre clinical databases in the United Kingdom found that only a few had realised their audit or research potential and that data quality varied greatly, particularly in terms of coverage and completeness (Black et al. 2004). A third of the database guardians didn't know their recruitment proportion. Funding is a key issue for such databases, which has a direct impact on their quality. Lastly, the types of unit of interest for performance monitoring, such as physician, general practice and hospital, must of course be identifiable in the database. This is usually the case with clinical databases though not always: the 2004 review found that it was possible for 95 of the 105 reviewed.

5.4 Accuracy of Administrative and Clinical Databases Compared

In general, there is good agreement between the two for what we might call administrative items, like dates and times of appointments and other contacts, and for basic demographic information, such as age and sex. Missing (incomplete) data afflict all types of databases. Common examples are tumour stage in cancer registries and comorbidity in administrative databases, though it's important to note that only in the latter case is the deficit not apparent from inspecting the record. If the secondary diagnosis fields are blank, it's either because the patient really did have no other diagnoses or it's because they did but those other diagnoses were not recorded for some reason. Most of the literature comparing the two types of databases concerns comorbidities and complications in hospitalised patients, partly because both sources rarely exist for other healthcare sectors. Before we consider this literature, we need to look at the evidence around case ascertainment, or "coverage" using the term in the earlier section on measuring data quality.

Our research group compared surgical databases with the national hospital administrative one for England for two specialties and found notable differences. For hospitals contributing to the Vascular Society of Great Britain and Ireland's National Vascular Database, there were twice as many procedures recorded within the national administrative data set as within the clinical database, though restricting the analysis to surgeons known to contribute to both narrowed the gap (Aylin et al. 2007a). A comparison of administrative data with the Association of Coloproctology of Great Britain and Ireland (ACPGBI) colorectal cancer database found the former to have 12% more cases overall and only modest correlation of hospital-level numbers and mortality rates (Garout et al. 2008). The 30-day mortality rate for those submitting to the national audit for colorectal surgery, the National Bowel Cancer Audit Programme (NBOCAP), was significantly lower than that for those units not submitting: 4.0% compared with 5.2% (Almoudaris et al. 2011). Even within the same hospital, not all surgeons submitted records to their society's database, whereas their hospital has to submit information for all operations to the administrative one. Even for surgeons who do submit, selective submission of their activity by omitting the "less interesting" cases, for example, also reduces the coverage and usefulness of the database.

Some studies have looked at medical rather than surgical conditions. A 2013 British study linked four databases to estimate the overlap and death rate following acute myocardial infarction (AMI): primary care electronic patient records from the Clinical Practice Research Datalink, hospital admissions from Hospital Episode Statistics (national administrative source), the national registry of acute coronary syndromes (Myocardial Ischaemia National Audit Project [MINAP]) and the national death registry (Herrett et al. 2013). Linkage was done deterministically using the date of birth, sex and the NHS number, which is unique to all UK patients and specific to healthcare. Ninety-six percent of patients with a valid NHS number were successfully matched. The crude incidence of AMI was underestimated by 25%–50% using one source compared with using all three sources excluding the death register; that is, each source missed a substantial proportion of cases. Some is inevitable, as HES and MINAP only include hospitalised patients, though this loss of coverage or case ascertainment is worrying. For the cases that did get captured by two or more databases, when using the registry as the gold standard, the PPV of the primary care records and of the hospital records were both estimated at 92%. This means that if the primary or secondary care record says the patient had an AMI, the patient usually had an AMI rather than some other main diagnosis. A PPV of over 90% for AMI in administrative data was also found by a systematic review (McCormick et al. 2014). Such low false positive rates help make such data suitable for cohort studies and trials as false positives dilute effects and reduce statistical power. For performance monitoring, a high PPV is also a good thing, though if you had to choose between a low false positive rate and a low false negative rate, it's less clear which is more important. If you want to compare performance between cardiac units or hospitals on their AMI mortality, for instance, is it less bad to include some non-AMI patients in the analysis (so a low specificity and/or low PPV) or miss out some genuine AMI patients (low sensitivity and/or low-negative predictive value)? The ideal is to have one accurate electronic record that is then augmented and propagated among other databases as required without the vast duplication of effort by audit staff, clinical coders and primary care staff. Meanwhile, the authors argue that more use should be made of database linkage, as is happening in initiatives such as Biobank in the United Kingdom (Biobank Protocol 2007) and the Million Veteran Program in the United States (U.S. Department of Veterans Affairs).

We now review the literature on the completeness of comorbidity recording in administrative databases compared with clinical ones. Comorbidity recording is important for reimbursement in a DRG-type system, health services research and risk adjustment of outcomes (Chapter 6), all of which makes it ripe for data quality assessments. A number of studies have shown that, compared with the original casenotes (charts), administrative data often under-record comorbidities. For example, Quan et al. (2002) examined 1996 data for around 1200 patients in Canada and found that the administrative data had a lower prevalence in 10 of the 17 Charlson index comorbidities, a higher prevalence in 3 and a similar prevalence in 4. The kappa values ranged from a high of 0.87 (near-perfect agreement between the two data sources) to a low of 0.34 (only fair to poor agreement): "agreement was therefore near-perfect for one variable, substantial for six, moderate for nine, and only fair for one variable". When 16 of these comorbidities were used to predict in-hospital mortality, the resulting pairs of odds ratios were similar for 10 and different for 6. A review of Canadian studies in this vein (Needham et al. 2005) concluded that the administrative data generally underestimated the incidence of individual comorbidities, especially for asymptomatic diseases, but that the Charlson score was less underestimated and the odds or rate ratios for mortality were similar to those from the clinical databases.

A Californian study on CABG (Parker et al. 2006) compared state discharge data with the registry from the California CABG Mortality Reporting Program, whose database and risk model closely follow those of the National Society of Thoracic Surgeons. They found kappa statistics for the comorbidities of interest ranging from 0.20 to 0.90, with a similarly large range in the PPVs. Diabetes and hypertension were far more reliably recorded in the state databases than heart block and dysrhythmia. A Swiss study of 3500 randomly selected patients (Januel et al. 2011) found that the introduction of ICD10 improved the sensitivity and PPV of many of the set of Charlson and Elixhauser comorbidities, though much under-recording by the administrative data and variation between conditions remained. The PPVs were at least moderate (mostly >50%) and exceeded 90% for many conditions such as hypertension, renal failure and liver disease. Sensitivities were mostly in the 30%–80% range, though anaemia and weight loss fared worse.

Some studies, including some of those just mentioned, went beyond reviewing data quality and estimated its impact on different kinds of analysis. The Canadian studies reviewed by Needham et al. (2005) found that, despite the administrative data under-recording comorbidities, the resulting odds or rate ratios for mortality were similar to those from the clinical databases. They were necessarily restricted to information that could be derived from both sources. As we have seen, registries and other clinical sources have the potential to capture comorbidity better but also go beyond and capture physiology and disease severity. They can also distinguish between pre-existing disease and complications: only some administrative sources can do this.

There is a growing body of work comparing the two sources in terms of their ability to predict and adjust for risk, much of it in cardiovascular conditions. The study of CABG mortality that used Californian state discharge abstracts and the registry (Parker et al. 2006) included a comparison of an established CABG clinical risk model against one derived from administrative data. The two models were compared in terms of statistical performance and hospital ratings. For some pre-operative risk factors such as the New York Heart Association classification and ejection fraction, no proxies in the state data could be derived, but for others some matching was possible. For example, the ICD code for obesity was used, and the creatinine level recorded in the registry was dichotomised into the binary flag of ICD's acute renal failure. In the two risk models, most odds ratios were in the same direction and of similar magnitude. The registry model showed slightly better discrimination ($c = 0.82$ versus 0.80) and slightly better calibration, though both models struggled a bit with the extremes of risk (see Section 6.8.4 for definitions of these model performance measures). In a study comparing the performance from mortality models from national administrative data with published figures from registries for CABG, abdominal aortic aneurysm repairs and colorectal excision for cancer, we found that the administrative models performed similarly in terms of discrimination to those from the registries (Aylin et al. 2007b).

Lastly, for our purposes of estimating performance at a unit level, it's not easy to predict without research to what extent the undoubtedly superior ability of clinical data to estimate an individual's likelihood of some outcome will translate into better risk adjustment of outcome measures when aggregated to the unit level. Happily, several studies have done this, largely from North America and partly prompted by the public reporting by the CMS of hospital mortality and readmission. We'll now describe them in a bit of detail.

Parker et al.'s 2006 Californian analysis showed a strong correlation between the two sets of hospital-level risk-adjusted mortality ratios (SMRs) for CABG (Spearman's rho = 0.87). SMRs were put into rating bands depending on how they compared with the state average: the two models agreed on the rating for 66 of the 75 hospitals (kappa = 0.67). The administrative model labelled 16 hospitals as outliers (11 better and 5 worse than expected), whereas the clinical model labelled 17 as outliers (13 better and 4 worse than expected). The models disagreed on the rating band for four low-mortality hospitals and five high-mortality hospitals. Most of these differences resulted from only very small changes in the SMRs. The authors noted that similar previous work on CABG found similar discrimination between the two data sources, though some of this apparently good performance by the administrative data was due to the inclusion of some complications that only the registries could distinguish from pre-existing conditions. Parker et al. were able to improve on some administrative models by making use of the ability to link patient records: around half of these CABG patients had been hospitalised in the state in the previous three years. The extra information they thereby derived affected the c statistic only a little but, as discussed in Chapter 6, it isn't right to conclude that this information is of no value. For example, their derived variable indicating prior open heart surgery had odds ratios of over 2.5 (p < 0.001). The authors conclude by asking the key question that we will come back to: "given the absence of clinical registries in most states, is [the administrative model] good enough?"

In the same year, Krumholz et al. (2006) published their validation of a model for 30-day mortality after admission for heart failure based on claims data subsequently used in CMS's public reporting program. They took the claims data from the Medicare Provider Analysis and Review (MEDPAR) files constructed a linked sample that contained both claims and medical chart abstracted data from the National Heart Care Project. The validation sample consisted of nearly 50,000 admissions and over 4,000 hospitals. Discrimination was better in the chart-based model (c = 0.78, maximum rescaled R^2 = 0.22 compared with c = 0.70, R^2 = 0.09 – higher values of c and R^2 are better); calibration was similar in both. A correlation between the two resulting sets of hospital-level SMRs was very high at rho = 0.95, and the median and maximum differences between the SMRs were negligible. As with their similar work on AMI readmission rates (Krumholz et al. 2011), they concluded that "the claims model was a very good surrogate for the medical record model".

A Canadian study (Austin and Tu 2006) calculated sets of SMRs derived from indirect standardisation and random effects models (see Chapter 7) using detailed clinical data collected on a random sample of nearly 12,000 AMI patients at 102 hospitals via chart review. These data were linked to the administrative discharge abstracts for the same hospital admission, and another set of SMRs were derived. A correlation between the two sets of SMRs was high (rho at least 0.87). Four and six hospitals were labelled as having mortality that was statistically significantly higher than expected by using administrative and clinical data, respectively, when model-based indirect standardisation was used. For random effects models, zero and two hospitals were labelled as having significantly higher mortality by using administrative and clinical data, respectively. This contrasts with the prevailing opinion that administrative data generate too many false positives compared with the more detailed source. We will return to the related debate over which type of SMR produces more outliers in Chapter 9. The different ways of obtaining SMRs will be described in Chapter 7.

Hammill et al. (2011) approached the question in a different way. They linked heart failure hospitalisations from the Get With The Guidelines–Heart Failure registry with Medicare claims data to estimate how much difference the extra variables from the registry made to the risk-adjustment model and to the hospital rankings. The claims-clinical mortality model had better fit than the claims-only mortality model (c = 0.76 and generalised

$R^2 = 0.17$ compared with $c = 0.72$ and $R^2 = 0.11$); no single variable accounted for the difference. The claims-only readmission model performed as well as the claims-clinical one ($c = 0.59$, $R^2 = 0.03$ compared with $c = 0.60$, $R^2 = 0.03$). They assessed consistency of hospital performance by how many hospitals moved in or out of the top and bottom 20% of hospitals. Rankings are inherently unstable (see Chapter 9), but dividing units into quartiles or quintiles is easy and still used in some public reporting schemes, which may explain why they're still used regularly in research. There were 308 hospitals in this study. Of the 68 ranked in the top quartile for mortality by at least one of the models, 54 were so ranked by both models; a further 7 were put into the middle band (middle 60%) only by the claims model and another 7 were put into that band only by the claims-clinical one. For readmission, there was much greater agreement, with the equivalent numbers being 58 in the top quartile by both models, and 2 and 2 more so labelled only by one of the models. Figures were similar when considering the worst performers for both outcomes.

Earlier work in cardiac surgery (Geraci et al. 2005) used the Department of Veterans Affairs Patient Treatment File and nurse-abstracted data from the Continuous Improvement in Cardiac Surgery Program. Discrimination was higher for the latter ($c = 0.76$ versus 0.70). One reason that the authors gave was that potential complications had been carefully filtered out at the cost of some potential comorbidities, but a bigger contribution is likely to be due to the lack of a flag to indicate the urgency of admission. This is generally one of the most potent risk predictors, which most other administrative databases have. From their results and previous work they cite, the authors recommended the addition of a very small number of variables to the claims database. The same team used similar data to compare SMRs for non-cardiac surgery (Gordon et al. 2005). They linked the VA Patient Treatment File to the National Surgical Quality Improvement Program database. Correlations between the clinical and administrative SMRs were 0.75, 0.83, 0.64, 0.78 and 0.86 for all surgery, general, orthopaedic, thoracic, and vascular surgery, respectively. These were fairly described as "moderately high". When compared with clinical models, administrative models identified outlier hospitals with a sensitivity of 73%, specificity of 89%, positive predictive value of 51% and negative predictive value of 96%. They conclude, "Our data suggest that risk adjustment of mortality using administrative data may be useful for screening hospitals for potential quality problems".

In summary, predictive models based on data which included clinical information were in general better able to estimate the risk of death (or other outcome) in individuals. However, once aggregated at an institutional level, models based purely on administrative data often compared favourably with clinical models when used for casemix adjustment and hospital comparisons, but it does depend on the databases in question. Table 5.1 summarises the main points regarding the continuing debate over the relative merits of clinical and administrative databases.

Sources of Data

TABLE 5.1

Comparison of Key Features of Clinical and Administrative Databases

Feature	Clinical Databases	Administrative Databases
Coverage, scope	Limited to small number of diseases or procedures.	Broad, often highly inclusive.
Regional coverage	Limited to participating centres, as usually voluntary.	Regional or national, allowing generalisability of analysis findings. Reporting is often mandated by law or incentivised through payments.
Unit and patient coverage	Voluntary participation can lead to bias against some types of unit and patient, resulting in lower average mortality rates than from administrative databases, for example.	Attempt to cover all units and patients, though private hospitals often not captured by the same database as public ones.
Size	Varies greatly.	Large, maximising statistical power to examine, e.g., rare events. Historical data often available to facilitate analysis of trends.
Diagnostic coding systems and associated issues	Varies, as can be tailored to the specialty.	Usually ICD, which has gaps, especially regarding disease severity and lab results. Present on admission (POA) information that distinguishes between comorbidities and complications is only captured in some countries.
Procedure coding systems and associated issues	Varies, as can be tailored to the specialty.	Often country specific. May not be a good fit to classifications used in clinical databases, impeding comparisons. Needs frequent updating due to new surgical techniques.
Data entry staff	Clinical (e.g., surgeons or research nurses), so recording of clinical items is likely to be more accurate than when done by administrative staff – but potential for gaming if data are used for performance monitoring.	Trained coders, adhering to coding guidelines. They are independent of clinical staff but can come under pressure from management to game data if used for performance monitoring.
Data quality	Perceived as high, which is a risky oversimplification. The best clinical databases will likely outperform the best administrative ones.	Perceived as low – also a risky oversimplification. Large size mitigates random but not systematic data errors.
Cost to maintain	High, both financially and in willpower, though often strong clinical engagement.	Low, due to pre-existing infrastructure for collection.
Key benefits in terms of data items	Often strong on physiological and other clinical information. Audit databases are designed for measuring care processes.	Often the only feasible source of financial information.

5.5 Incident Reports and Other Ways to Capture Safety Events

In this section, we will cover voluntary and mandatory reporting systems of safety events that have already occurred and will also briefly touch on prospective methods for detecting system weaknesses and defects that could lead to safety events before they happen. Although safety is just one dimension of quality, it's so complex that it's often considered separately and has a large number of diverse methods devoted to measuring it. We've already seen how administrative databases have been used to estimate event rates and flag cases for more detailed review, but they were not designed to capture such information – unlike patient safety event reporting systems. These systems are now very common in hospitals in developed countries and increasingly in LMICs too. When safety events of whatever kind are reported by those involved in them, this is voluntary and is a form of passive surveillance. In contrast, events can be sought by active surveillance using chart review, for instance, using tools such as IHI's Global Trigger Tool, using administrative databases or incident report databases, or using ethnographic approaches that use direct observation of staff at work. The last of these have the extra benefit of being able to identify the unintended consequences of safety policies.

In addition to active or passive forms of surveillance, attempts to measure patient safety can be divided into prospective and retrospective methods. Prospective methods try to detect system defects before errors happen. Several approaches are established for doing this. In addition to direct observation mentioned earlier, there are formal safety culture surveys (see Colla et al. 2005, for a review of nine of them) and more informal methods such as executive walk rounds. Failure mode and effect analysis (FMEA) is a common prospective hazard analysis method used in many industries and assesses the risk of error within a particular process. First, an experienced, multidisciplinary team map out the process's steps. Next, in the failure modes part, the team considers how each step can go wrong. Then, in the effect analysis part, they estimate the probability that each error will be detected before causing harm and the impact of the error if it actually occurs. These are combined into a criticality index for each step, a rough estimate of the magnitude of the hazard that it poses, so that priority can be given to improving those steps with the highest criticality indices. One drawback of this is that different teams can assess the hazard associated with each step quite differently, another example of the problem of interrater reliability. Another more qualitative approach is SWIFT, "structured what-if technique", which can be used either as an adjunct to FMEA or by itself. Potts et al. (2014) compared the two in a community-based anticoagulation clinic with previous assessments of risk done by root cause analysis and staff interviews. They found limited overlap between the methods in hazards that each identified and concluded, as is usual in patient safety studies, that multiple data sources need to be used.

Regarding retrospective approaches to measuring safety events, one source of information that staff will likely use is the incident reporting database. In addition to those in individual institutions, some countries have national systems. The National Patient Safety Agency (NPSA) in the United Kingdom set up the National Reporting and Learning System (NRLS) in 2004. It uses the definition of a patient safety incident as "any unintended or unexpected incident that could have or did lead to harm for one or more patients receiving NHS-funded healthcare" and so also includes near misses. It includes all types of safety incident, unlike other systems in the United Kingdom such as the Medicines and Healthcare products Regulatory Agency "Yellow Card Scheme" for side effects from medicines, vaccines, medical devices and herbal or complementary remedies. The Yellow Card Scheme acts as an early warning system for the identification of previously unrecognised adverse reactions: anyone can submit a report via https://yellowcard.mhra.gov.uk/. The public can also submit to the NRLS, though the majority of incidents are submitted electronically from local NHS organisation risk management systems, and the reporters are usually nurses. The agency gives guidance to institutions on how to check data quality. Since January 2009, NRLS summary data on the type of incident, care setting and degree of harm have been published each quarter. Information is also published by NHS institution, which covers not just hospitals but also community and ambulance services. The types of incidents currently reported to the NRLS are shown in Box 5.1.

BOX 5.1 TYPES OF INCIDENTS REPORTED TO THE NATIONAL REPORTING AND LEARNING SYSTEM IN ENGLAND AND WALES

1. Access, admission, transfer, discharge (including missing patients)
2. Clinical assessment (including diagnosis, scans, tests, assessments)
3. Consent, communication, confidentiality
4. Disruptive, aggressive behaviour
5. Documentation (including records, identification)
6. Infection control incident
7. Implementation of care and ongoing monitoring/review
8. Infrastructure (including staffing, facilities, environment)
9. Medical devices/equipment
10. Medication
11. Patient abuse (by staff/third party)
12. Patient accidents
13. Self-harming behaviour
14. Treatment, procedure
15. Other

The number of incidents is huge. For example, 1,637,260 were reported across England and Wales between October and December 2014, of which 73% came from hospitals. Overall, 70% were assessed as no harm and 24% low harm, with 3,629 (1%) as severe harm or death. In terms of the type of incident, the top three overall were patient accident (21%), implementation of care and ongoing monitoring/review (12%) and medication incidents (11%), but these varied a lot by setting. For ambulances, a quarter of incidents were in the category of access, admission, transfer and discharge, and, of the 5000 or so incidents reported in general practice, the most common was around implementation of care and ongoing monitoring/review. Studies of electronic hospital event reporting systems generally show that medication errors and patient falls are among the most frequently reported events (Milch et al. 2006).

The AHRQ gives these four key attributes of an effective reporting system:

1. A supportive environment for event reporting
2. Reporting by a broad range of personnel
3. Timely dissemination of event summaries
4. Structured process for reviewing reports and developing action plans

There are several limitations to voluntary systems, including the lack of input from physicians and the overrepresentation of minor incidents and near misses (if the system allows their capture). The reporter's identity is usually known in voluntary systems, but there's legal protection unless there has been professional misconduct or criminal acts. However, bigger drivers of low reporting by both physicians and allied health professionals than concern over personal repercussions include lack of time and failure to receive feedback after reporting an event (Evans et al. 2006; Farley et al. 2008). Due to all these limitations, the NPSA advise against the tempting interpretation of a high incident rate in one hospital compared with another as reflecting poorer safety standards. They still suggest that having a high number of reported incidents reflects a more mature and healthy safety culture, that is, high equals good.

Systems within countries that lack a national system vary, as do systems between countries. The World Alliance for Patient Safety Drafting Group of the World Health Organization therefore created an internationally agreed conceptual framework for the International Classification of Patient Safety to facilitate the description, measurement, monitoring, analysis, comparison and interpretation of patient safety data (Sherman et al. 2009). Based on this conceptual framework, the WHO developed a template, the Minimal Information Model, as a minimum set of common data categories to facilitate reporting and comparison of units. The Minimal Information Model comprises details of the incident (patient, time, location, agent[s] involved), type, outcomes, resulting actions and reporter.

This discussion has concerned only voluntary reporting, but some countries (also) have elements of mandatory reporting. In the United States, such systems are usually enacted under State law, such as Pennsylvania's Medical Care Availability and Reduction of Error (MCARE) Act of 2002. These generally cover sentinel events, such as specific errors or those leading to serious patient injury or death. A number of European countries have mandatory reporting for healthcare professionals and a few for healthcare institutions (European Commission 2014). In Sweden, for example, according to the 2011 Patient Safety Act, providers must notify the Health and Social Care Inspectorate serious adverse events that resulted from or could have resulted in a serious incident. Reporting to other systems is voluntary.

5.6 Surveys

If you want someone's opinion on something, you need to ask them for it or give them a forum to express it (see later for social media). Asking about patient experience is recognised as the key way to assess to what extent the healthcare system and its providers are patient-centred, an IoM domain of quality, though it can also capture information about other domains. A number of countries now have national survey programs to capture patient opinions on their healthcare and healthcare provider. Formed in 2000, the Picker Institute designed and delivered the first national NHS patient experience survey in England and runs similar programs in other care settings and countries. In the United States, Press-Ganey has been measuring patient satisfaction for twice as long. They sample patients in different ways, such as what they call "census-based" (via email, phone and regular mail) and visit- or encounter-based. Agency for Healthcare Research and Quality Consumer Assessment of Health Providers and Systems (CAHPS) surveys cover a range of settings. For example, the CAHPS Clinician and Group questionnaire core items ask about getting timely appointments, how well providers communicate with patients, whether the staff are helpful, courteous and respectful and ask patients for a rating of the provider. The results are linked to Medicare reimbursement. In the United States, the public get asked a lot about their health and healthcare by providers, health plans, employers and government agencies. In Australian hospitals, there is a wide variety of standardised and locally developed surveys. Most of them capture information on things such as waiting times, access, information sharing and communication, the physical environment, overall satisfaction, patient involvement in care decisions, privacy/respect/dignity and coordination and continuity of care (Board 2012). Surveys are also used to collect patient-reported outcome measures. See Chapter 4 for more on PREMs and PROMs.

Some countries also send surveys to staff. In England, the annual NHS Staff Survey began in 2003 and is coordinated by the Picker Institute on behalf of NHS England. According to their website, it "is recognised as an important way of ensuring that the views of staff working in the NHS inform local improvements and input into local and national assessments of quality, safety, and delivery of the NHS Constitution". The 2014 survey was sent to a record 624,000 of the 1.2 million people that the NHS employs and asked about their perceptions of their organisation and job satisfaction. Sixty-seven percent of staff said that patient care was their organisation's top priority; less than two-thirds said they would be happy with the standard of care provided by their organisation if a friend or relative needed treatment there. Only 56% of staff would recommend their organisation as a place to work, and only 41% felt that their organisation valued their work. Just 29% thought there were enough staff for them to do their jobs properly. The survey also asks about factors, including satisfaction with pay, levels of bullying and harassment, and sickness.

Survey designers and implementers need to be aware of language and cultural differences. Translation into languages other than a country's official one(s) is costly but important, particularly as minority communities are frequently affected by difficulties in accessing good care. Sometimes, more than just translation of the words is required. Written questionnaires presume literacy. In hospitals in Australia's Northern Territory, for instance, meaningful pictures and symbols are incorporated within surveys. Cultural differences can affect how people respond; for example, they affect how likely people are to use the extremes of Likert-type rating scales. They are also known to affect satisfaction ratings, which are driven by people's preferences and expectations. The move away from fixed landline to mobile telephones, especially among younger people, has implications for how best to contact people. It is also important to consider survey mode effects, which describe the different responses people can give depending on the manner in which they are asked, for example, face to face, by phone or in writing. In a study of CAHPS hospital survey scores, 27,000 patients were randomised to receive the survey through one of the four possible modes: mail only, telephone only, active interactive voice response (IVR) that involves automation and responding using the telephone keypad, and mixed mode (mail first, then telephone follow-up if no response to the mail invitation). In general, patients gave more positive responses regarding their experience in hospital if they completed the survey by phone (either telephone or active IVR modes) than by mail (Elliott et al. 2009). These mode effects were certainly big enough to alter a hospital's apparent performance rating. The study also encountered the well-known effects from psychology of social desirability bias – people giving what they think is a more acceptable answer, which happens more often in face-to-face interviews and least often in written ones – and the "recency effect", meaning a tendency to pick the last option within a list with an auditory rather than visual presentation, that is, to pick the most recent option that the respondent has heard.

There are now a number of validated survey instruments for assessing patient and staff experience. The usual issues need to be weighed up when considering if and how to use this type of information: sampling frame, response rate and overall fitness for purpose. Does it ask the questions you need of the right people at the right time, and can you trust the information you get back?

5.7 Other Sources

The sources we've discussed up to this point all involve the active collection or extraction of information from patients. Increasingly, though, patients are volunteering information, sometimes with relish. The explosion of social media and websites devoted to soliciting opinions on healthcare shouldn't be overlooked as being of potential value to those seeking to evaluate healthcare providers and indeed the whole healthcare system. Surveys inevitably ask specific and limited questions and are done infrequently because of the costs of administering them, but people can give themselves free rein online. Examples of patient websites that allow users to leave feedback include iwantgreatcare.com, ratemds.com, ratemyphysician, healthgrades.com, ratemyhospital, and NHS Choices. Some ask for a star rating in the manner of TripAdvisor, whereas others allow free text comments. For example, if you go to healthgrades.com, choose a city and click on a procedure such as knee replacement, you'll be given a list of physicians in that city who do knee replacements, put into the descending order of overall satisfaction rating. For each physician's office and staff, a star rating out of five is given for ease of scheduling urgent appointments; office environment, cleanliness, comfort, and so on; staff friendliness and courteousness; total wait time (waiting and exam rooms). Below that are ratings under "Experience with Dr X": level of trust in provider's decisions; how well the provider explains medical condition(s); and how well the provider listens and answers questions and spends appropriate amount of time with patients. The number of responses, that is, people who have filled in the online questionnaire, is given. This kind of information is relatively easy to analyse once collected, although see Chapter 8 on composite measures for the pros and cons of producing an average satisfaction score.

Non-response bias will naturally come to mind with all these ratings websites. The potential use of patient opinions when culled from social media platforms like Twitter or sites that allow free text comments such as NHS Choices or ratemds.com has the issue of (non-)response bias plus the difficulties of how to convert the free text, often with spelling errors, into useable information. This is a fairly new field but already there have been some attempts to try to make sense of it. In the United States in 2010, 11% of adults had consulted online reviews of hospitals or other medical facilities (Fox 2011). A 2014 mixed methods study comprised a quantitative analysis of all 198,499 tweets sent to English hospitals over a year and a qualitative directed content analysis of 1000 random tweets (Greaves et al. 2014). It found that 11% of tweets to hospitals contained information about care quality, most frequently regarding patient experience (8%). Comments on effectiveness or safety of care were less common (3%). Seventy-seven percent of tweets about care quality were positive in tone. The authors concluded that the tweets were of potential use for quality improvement but, especially given the lack of correlation with established measures of quality, not really for performance monitoring. A sentiment analysis of free text comments left on NHS Choices found a good correlation between patient's views of the hospital with those expressed in the paper-based NHS national inpatient survey. This was particularly true for the questions around cleanliness, being treated with dignity and overall recommendations of the hospital (Greaves et al. 2013). The authors acknowledge this as early work, as the methods for analysing free text are complicated and evolving. Given the potential for non-response bias in the paper-based formal surveys, we should be cautious about rushing to use this unstructured information for performance monitoring – but we should keep an eye on the field as it advances.

5.8 Other Issues Concerning Data Sources

With patient-level confidential data, there are governance and ethical issues to be considered. Many countries have laws such as the UK Data Protection Act 1998 to safeguard personal data. When presenting results of analyses, users need to be careful not to publish information that is potentially disclosive. A common example is the suppression of low counts for a particular disease in a particular geographical area, such as those between 1 and 4.

The main ethical issue is the so-called secondary use of data, such as when electronic health records formed during the clinical management of a patient are extracted, often anonymised or pseudonymised, and passed on to third parties such as academic institutions for research. The data were collected for one purpose, to which the patient has given their consent, and are then used for another purpose, for which he or she has not given consent; for day-to-day clinical care, the patient generally gives implicit consent for their personal information to be stored and used by staff directly involved in treating them, whereas for clinical trials and registries, the consent would be explicitly requested. The degree to which society deems secondary use acceptable varies by country. Where societies are uneasy about this, individual consent can nonetheless be waived by oversight bodies on a case-by-case basis.

This happens only if those bodies deem that the secondary use, for example, the particular research project, is in the public interest and that it would not be feasible to try to get consent for the patients whose data would be used in that research project. Obtaining access to administrative databases can be significantly delayed by various steps in the approval process.

5.9 Conclusion

In summary, high-income countries typically have a welter of databases for use in performance monitoring, each with their own strengths and limitations and each with their own particular but incomplete story to tell about the healthcare system.

6
Risk-Adjustment Principles and Methods

Aims of This Chapter
- To distinguish between risk adjustment and risk prediction
- To set out the need for risk adjustment and main principles behind building a risk-adjustment model
- To describe how to assess whether the adjustment has been successful

In this chapter, we will assume that we have chosen and defined our measures and data sources in alignment with the purpose of our analysis. Regarding the purpose of risk adjustment, Iezzoni (2013) lists as follows: to set payment levels for individual patients, to encourage providers to accept high-risk patients, to compare efficiency and cost across plans/providers, to give performance report cards for the public and to allow benchmarking for quality improvement. The latter three will be our focus. We will outline what risk adjustment is, why and when it's needed, some alternatives to it (these will crop up often in the discussion), some principles for doing it and how we might assess how well we've done it. The next chapter will cover how the risk-adjustment model leads to a risk-adjusted outcome for each unit: there's a surprisingly large number of ways this can be done, and they don't all give the same answers.

This chapter will only consider the derivation of cross-sectional outputs such as hospital report cards summarising the performance for a given year. Chapter 12 discusses before versus after comparisons using time series data and also covers other means of dealing with confounding such as propensity score matching and instrumental variables analysis, which might belong here but, as they (so far) are more usually applied to evaluations of interventions, we've left them until that later chapter.

6.1 Risk Adjustment and Risk Prediction

Risk adjustment and risk prediction have many similarities but also some key differences. In risk prediction, we typically want to predict whether a given patient will go on and have some given outcome. An example is the Modified Early Warning Score in acute medicine, of which there are several in use worldwide, which aims to help ward staff spot which patients are at high risk of imminent deterioration to enable timely life-saving interventions such as calling the emergency medical team or transferring to intensive care. Such a score will ideally be calculable at the bedside, perhaps in the physician's head, using just a few variables (models with minimal factors included are said to be parsimonious). Measures of statistical performance of screening tests such as sensitivity and positive predictive value can be used to judge how good the risk-prediction model is compared with what happened to the patients in reality. As the consequence for not responding to a high Modified Early Warning Score value could be death, one could argue that the score needs a high sensitivity to be of value, at the cost of a number of false alarms – we will discuss the value of such statistical measures in different scenarios later in this chapter. Risk prediction for surgical patients could also take into account the hospital at which the operation will be performed and/or even the surgeon doing the operation. From the patient's perspective, it's more useful to know the outcome expected for someone of their age and disease with that particular surgeon supported by his or her particular team than the outcome expected for some of their age and disease nationally. It makes a difference if the surgeon is an expert at a centre of excellence or if the surgeon is a relative newcomer at a struggling hospital. Finally, a risk-prediction model may work very well even if it includes biased coefficients and omits variables with strong associations with the outcome (Copas 1983). Small and simple are best.

In contrast, the goal of risk adjustment is to control for casemix differences between units. The emphasis is not on parsimony or interpretability – in theory, it doesn't matter if we use 10 coefficients to describe the relation between age and mortality if that's how age affects the chances of death in reality. What matters is that hospitals or other units being compared can't complain that they're being penalised due to having older patients than other units. Models can therefore include a large number of factors combined using a complex algorithm, though it's true that risk-prediction algorithms can also be complicated if they don't need calculating by hand. Risk-adjustment models typically combine information across a number of providers, need to watch for biased coefficients for patient factors that differ between units, and, for reasons we will explore shortly, rarely adjust for surgeon-level or hospital-level factors.

A last distinction between the two is that greater accuracy is needed when predicting risk for an individual patient than for accounting for differences in casemix between units. Risk estimates are more accurate for populations of similar patients than they are for any individual patient. This is because of random variation and the fact that we never have information on all the predictors. We can intuitively see that this difference in accuracy is because a number of differences between patients tend to cancel each other out when considering a group of patients together, as is the case for us with performance monitoring at the level of healthcare unit.

6.2 When and Why Should We Adjust for Risk?

A common response by a provider on seeing that their outcome rates are worse than the benchmark is, "That's because our patients are sicker". One task of risk adjustment is to overcome this objection and take account of patient risk factors for that outcome. There are other tricks that can be used separately or in conjunction in order to make the comparisons fairer, which will be considered at the end of this chapter, particularly risk stratification.

Thus far, we've consistently talked about outcome measures. Donabedian's classic triad also included structure and process measures. In general, risk adjustment is only needed when comparing outcomes between units, as casemix can have large impacts on the chance of a given outcome and is generally beyond the control of the unit. When comparing units in terms of structure measures, such as staffing levels or equipment, no adjustment for patient risk is needed or indeed possible as these measures involve unit-level data. What will be important, however, is an appropriate benchmark. A village clinic will obviously compare poorly on MRI scanners with a teaching hospital. For process measures, the picture is a little muddier than those who advocate focusing exclusively on processes over outcomes would have you believe. It is certainly true that risk adjustment is far less relevant for process measures. Patients should be given antibiotic and venous thromboembolism (VTE) prophylaxis before many kinds of surgery. Patients with heart failure should be given beta blockers and angiotensin converting enzyme (ACE) inhibitors. There are exceptions to such standards, particularly if the patient has contraindications, and for some processes patient preferences will come into play. With careful definition of the measure's denominator and of course available data, we can get round the problem of contraindications. Then the inverse care law strikes. Many studies of variations in process achievement between units have shown that such variations can be explained by differences in casemix, such as age, sex, ethnic group and socioeconomic status (SES) (Rubin et al. 2001). These reflect in part the difficulties in obtaining consent or cooperation, for instance, due to language differences or cognitive difficulties. If, following an audit, your unit is found to do relatively poorly on some process measure, then analysis by patient subgroup can help to elucidate where the shortcomings are. Choice of a more appropriate benchmark such as your peer group will help too – if the unit down the road, which serves the same population, performs better than your unit, then they may have something from which you can learn.

6.3 Alternatives to Risk Adjustment

As discussed in Chapter 9, the choice of benchmark is important. The grumbling unit's patients may be sicker than those at an average general hospital but be similar to those at a tertiary centre – in that case, they should be benchmarked with other tertiary centres. Another strategy is to stratify the comparative analysis by patient subgroup; careful choice of inclusion and exclusion criteria can also help make the set of patients more homogeneous. We therefore have two key strategies for accounting for risk: include and adjust, or exclude and stratify. Stratification is a vital tool for reasons that will be explained during the rest of this chapter. It has its limits, as splitting the results by too many patient groups (strata) can become unwieldy and steadily loses statistical power to detect performance differences.

6.4 What Factors Should We Adjust For?

6.4.1 Factors Not under the Control of the Provider

A standard principle of risk adjustment is to only account for factors that are not under the control of the provider. A surgeon performing a heart bypass with valve repair knows they will have a tougher job with higher risks for the patient than when they perform an isolated, first-time coronary artery bypass graft (CABG). It would be unfair not to take into account the extra risks inherent in the former procedure when comparing surgeons' performance. Indeed, in this example, it's usual for only first-time, isolated CABGs to be included in the comparisons, with the subgroup involving valve repairs compared separately: this is an example of "risk stratification". Assuming that the valve repair is clinically necessary, that is, there is little element of surgeon choice in whether to repair it or not, then our analysis should account for the extra-operative complexity. Similarly, if the patient is relatively frail or has certain diseases that

Risk-Adjustment Principles and Methods

increase (or decrease) their risk of the outcome of interest, we should also account for these because they are outside the control of the surgeon. Clearly, the provider can't (yet) make people younger or (yet) cure their dementia, but they can potentially change their staffing levels and intensive therapy unit (ITU) bed provision. We should therefore not adjust for the latter factors.

The question of which factors are under the control of the provider is actually rather nuanced. Age is the classic confounder in epidemiology and classic risk factor in medical research. It's associated with all manner of adverse events and outcomes. It would therefore seem natural to adjust for it whenever it predicts the outcome. One might also include it a priori even if the analysis doesn't show an association with your particular data set for some reason. However, consider Figure 6.1. It compares the death rates for elective (planned) paediatric cardiac surgery at the Bristol Royal Infirmary with those for England's other specialist centres for these procedures in the early 1990s. The x-axis gives the age in months at operation.

Clinical guidelines indicate that these operations should be done quite soon after birth, reflected in the light peaks in the first few months for the non-Bristol centres. At Bristol, however, the handful of cases operated on early all died, so the surgeons delayed the procedure for most cases until the child was older and their heart was bigger. Their high death rate – shown with both administrative data as in the graph and with the surgical society's

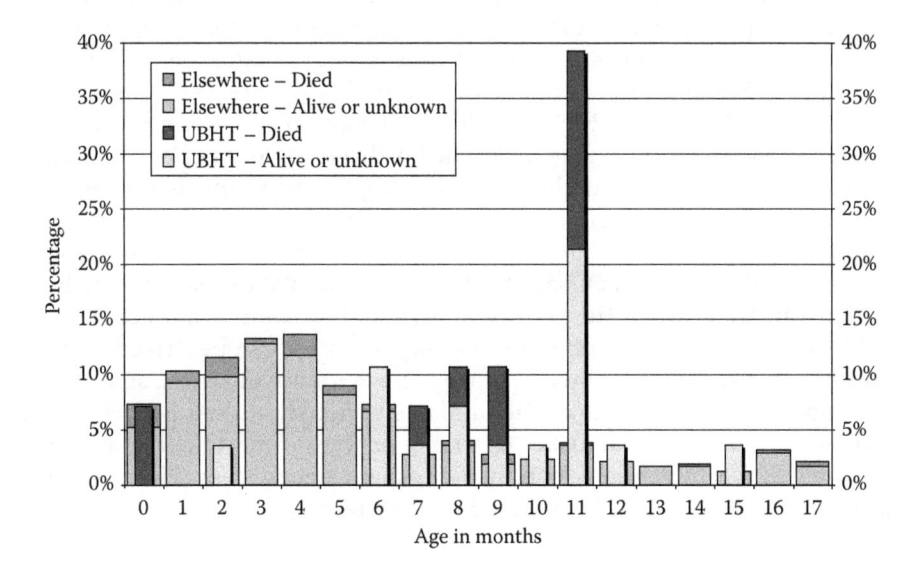

FIGURE 6.1
Comparison of percentage of AVSD operations by age at operation (in months) between Bristol (UBHT) and elsewhere in England 1991/1992–1994/1995. (Taken from Aylin, P. et al., Analysis of Hospital Episode Statistics for the Bristol Royal Infirmary Inquiry, Division Primary Care & Population Health Sciences, Imperial College School of Medicine, London, UK, October 1999.)

own database – and the reasons behind it became the focus of a public inquiry (Kennedy 2001) that led to a greater interest in the United Kingdom in the use of administrative data for performance monitoring, as we saw in Chapter 2. As these operations were planned, the age of the patient at surgery *was* under the surgeon's control. Adjustment for age in this case would have obscured some of the differences in the quality of care that we wanted to show. Apart from thinking through the problem, the other learning point here as regards modelling is that it's advisable to check the distribution of each potential risk factor by unit before beginning the modelling. This point will appear again later.

6.4.2 Proxies Such as Age and Socioeconomic Status

Age is a proxy for biological fitness or health status and, like all proxies, it's imperfect in the absence of what we really want to measure but nonetheless often performs well. Another common one is SES or deprivation. There is much debate over whether this should be included in risk-adjustment models. On the one hand, it captures factors such as lifestyle, housing, education, language barriers and environment pollution that all increase the patient's risk of poor health and/or getting poorer outcomes following healthcare. This argues in favour of including it, assuming it can be measured well for the intended purpose. On the other hand, the inverse care law argues against its inclusion: it's up to each provider to meet their patients' needs whatever their socioeconomic circumstances. Equity of access, for instance, is one of the Institute of Medicine's six dimensions of quality of care. What should we do? Jha and Zaslavsky (2014) say, "The decision to adjust for SES is situation dependent and may be different for three key uses of performance scores: informing patient choice, motivating quality improvement, and determining financial incentives". We can try to weigh up the relative importance in our situation of the two opposing forces of casemix and care inequalities and then satisfy the major one. We can stratify by SES, which can be a powerful way of showing that some units are able to achieve good results despite the challenges. Or we can be driven by face validity and stakeholder buy-in. If we have evidence or strong suspicions that no one is going to take the analysis seriously unless SES is in the model, even if it "overadjusts" for risk – or they won't take the analysis seriously if it *is* included – then that gives us a strong steer. The customer is always right (except of course when they're not).

6.4.3 Comorbidity

Comorbidity is commonly included in risk-adjustment models. A number of comorbidity indices have been developed, with the Charlson and its variants and the Elixhauser being the most commonly used

(Charlson et al. 1987; Elixhauser et al. 1998). Despite being developed on a fairly small set of records for breast cancer patients to predict mortality more than 25 years ago (and using weights based on odds ratios rather than the more mathematically correct log odds ratios), the Charlson index has been shown to work well for a range of outcomes. The later Elixhauser index was developed as a general purpose comorbidity index with administrative databases without present on admission (POA) information. It therefore doesn't include acute myocardial infarction (AMI) or stroke, which can both occur as complications after admission. Our group's systematic review of studies that compared two or more such indices found that Elixhauser tended to discriminate better for mortality than Charlson (Sharabiani et al. 2012). We and others have found it useful to include interaction terms between age and comorbidity and, if using Elixhauser, to add dementia (from Charlson) to its set. If possible, derive the weights from the database used in the modelling for whichever index is chosen rather than rely on published weights. These weights should also if possible be tailored to both the patient group and the outcome of interest: the relative importance of diabetes, for example, in predicting mortality following CABG is not necessarily the same as it is in predicting readmission following hospitalisation for heart failure. For a description of other comorbidity adjusters, see our review (Sharabiani et al. 2012).

Some administrative hospital databases such as in Canada, Australia and the Centers for Medicare & Medicaid Services (CMS) in the United States now include POA flags for each secondary diagnosis; many U.S. states are following suit in their all-payer inpatient discharge data sets. When the United States transitioned to International Statistical Classification of Diseases and Related Health Problems (ICD)-10-CM on 1 October 2015, it reduced the burden on coders via an exempt category when the diagnosis-timing status is irrelevant or inherent to the code. Research has shown that, given good documentation, rates of missing data for the diagnosis timing are generally now very low (Research Triangle 2011). Coders can often reliably make the distinction between POA and not (Sundararajan et al. 2015), though a study from California found the reliability for hip fracture patients to be much higher than that for community-acquired pneumonia (Haas et al. 2010). Some U.S. experts we've spoken to have doubts over the consistency of the POA flag across hospitals in CMS data. A study of patient discharge data in 48 Californian hospitals in 2005 found that overall reliability was reasonable but that, despite over a decade of experience of POA reporting in the state database, sizeable variations existed between hospitals (Goldman et al. 2011). Using this information in risk adjustment has implications. First, the odds ratios decrease for in-hospital mortality associated with the presence of some comorbidities, presumably because these comorbidities confer greater risk when they arise acutely during a hospitalisation than when they are chronic. A second consequence is that one can include in the risk models extra diagnosis covariates that would otherwise be excluded due to the uncertainty over whether they occurred in hospital. This will improve risk model performance. As an aside, due to reduced confounding, this may also shrink associations between hospital characteristics such as surgical volume and adjusted outcomes towards the null (Stukenborg et al. 2005). The WHO Quality and Safety Topic Advisory Group for ICD11 concluded, "Despite imperfect validity, the evidence to date indicates that a diagnosis-timing flag improves the ability of routinely collected, coded hospital discharge data to support outcomes research and the development of quality and safety indicators" (Sundararajan et al. 2015).

The set of comorbidities returned by patients when asked in interviews or questionnaires don't necessarily match those from their notes or charts. Veterans affairs (VA) clinic patients correctly reported their chronic obstructive pulmonary disease (97%) and coronary artery disease (95%) (Fan et al. 2002). The longitudinal Canadian National Population Health Survey found that 90% of patients recalled their major depression if it occurred in the previous year but that this dropped to 41% if it had occurred 10 years earlier (Patten et al. 2011). Another prospective study, this time of around 3000 patients who survived prostate cancer, assessed the consistency of patient-reported comorbidities across three time points (Klabunde et al. 2005). Response consistencies of 92% or better were obtained for 11 of the 12 comorbidities that were looked at. Regression analyses of the four most prevalent comorbid conditions (arthritis, diabetes, hypertension and depression) showed that consistency was higher for those respondents who reported that they were limited by, or taking prescription medicine for, the condition than those who weren't.

The data source can have other influences on data quality and apparent comorbidity levels. The electronic medical record (EMR) and paper notes or charts reflect physician diagnostic and record-keeping abilities, which in turn correlate with quality of care. Blood pressure (BP) readings taken by humans are subject to digit preference, in which the measurer rounds to the nearest 5 or 10; there is also the potential for gaming here if exceeding some BP threshold triggers either payment or penalty. Comorbidities are particularly potentially vulnerable to "upcoding" if greater recorded prevalence increases the predicted risk in a pay for performance (P4P) scheme just as it does in Diagnosis-Related Group (DRG)-based payment systems.

6.4.4 Disease Severity

It makes sense to try to adjust for disease severity. Unfortunately, this is usually hard to establish from administrative databases and so gives clinical databases an important advantage. ICD does not in general give much detail.

While there are separate codes for asthma and status asthmaticus, for example, there aren't any to give peak flow readings. If a subjective assessment is given, then bear in mind that the patient's rating tends to be more severe than the doctor's and will be influenced by the patient's mood and culture. An analysis of 233 patients with rheumatoid arthritis compared patient and physician rating of rheumatoid arthritis severity (Barton et al. 2010). Nearly a third of the patients differed from their physicians to a meaningful degree (defined as >25 mm on a 100 mm visual analogue scale) in their assessment of global disease severity. Such a difference was more common in those with more depressive symptoms and less common with a greater swollen joint count and in speakers of Cantonese or Mandarin.

6.5 Selecting an Initial Set of Candidate Variables

Having arrived at a long list of potential risk factors to put into the model, from reviewing the literature, a priori knowledge, consulting expert panels and of course seeing what's available in your data set, you need to consider whether to reduce the long list to a more manageable one and how each factor influences the outcome (e.g., linearly or some U-shaped relation). Using an expert panel to select risk factors will increase user buy-in, as done by the Medicare Mortality Predictor System in the 1980s and the VA Surgical Risk Study. The reliability of recording of factors is important. An example is the use of palliative care services during a hospitalisation. This information can be taken from either the specialty code or from secondary diagnosis codes (in ICD10 this is Z515). Unpublished analysis with English data showed that the proportion of deaths with palliative care recorded ranged from about 1% to over 50% by hospital in 2011/2012. The use of palliative care is one factor used in the model underpinning the publicly available hospital standardised mortality rate (HSMR) and can markedly affect the predicted number of deaths (Bottle et al. 2011). While it is possible that coding guidelines were being bent at some places to produce a more favourable HSMR, it's certainly true that those coding guidelines regarding the recording of the involvement of palliative care teams were confusing (and still are at the time of writing). For that reason, the newer NHS overall hospital mortality measure, the Summary Hospital Mortality Indicator, did not include adjustment for palliative care as the flag is considered to add more noise than it removes (HSCIC 2013).

All this is done before we do any modelling. In Section 6.8.1, we address how to reduce the set of candidate variables from this process to the final set of predictors. Before we can get down to the modelling in earnest, though, we need to consider two more data issues: missing or extreme values and the timing of the risk factor.

6.6 Dealing with Missing and Extreme Values

With some databases and fields, missing values present a problem that must be dealt with before modelling can begin. Missing values are a nuisance, but they can represent any of various patterns, and if they're sufficiently common it's important to understand why they're missing. They can mean simply "normal", for example, term birth in the gestational age field. They can be missing randomly – so-called missing completely at random (MCAR) is one of the ideal scenarios for modellers as it doesn't cause bias in a unit's estimated performance. MCAR is sometimes called uniform non-response in sample survey terminology and occurs if the probability of an observation being missing does not depend on observed or unobserved measurements, that is, it doesn't correlate with either the outcome or any of the potential predictors.

Data are said to be missing at random (MAR) if the probability of an observation being missing doesn't depend on the outcome but does depend on other risk factors. What looks at first glance as if it should be MCAR can turn out to be only missing at random. For example, a patient in a clinical trial may be lost to follow-up after falling under a bus. If it's a psychiatric trial, this apparent accident might in fact be due to poor response to treatment. Sensitivity analyses can assess the potential impact of this on the outcome (and on the treatment effect in clinical trials).

Another possibility is that missing data can sometimes correlate with the outcome. For instance, if the patient died soon after admission to hospital then they won't have any test results. This is the most problematic and is often handled with sensitivity analyses.

There are three main ways to deal with missing data of any kind:

1. Drop the affected cases and run a complete-case analysis
2. Drop the affected risk factor
3. Impute values

If a risk factor is particularly plagued by missing values and it's not a key predictor, then it's probably best to drop it from the model. If it is particularly important, but it's only missing for a few cases, then it may be best to drop those cases. This is the most common approach and is fine with data that are MCAR. If the risk factor is important and missing for a substantial proportion such as >10%, then imputation will be necessary. This is easier for continuous variables than for categorical ones. Multiple imputation is to be preferred to single imputation. The latter involves replacing the missing value with, for example, the overall mean, the patient's first value (called first or baseline observation carried forwards) or the patient's previous value (called last observation carried forwards). It used to be very common but doesn't account for the inherent uncertainty over whether you've picked an "appropriate" value. It's largely been superseded by the more complicated multiple imputation (Rubin 1987). In this method, a number of copies of the data set are made. Each imputed data set is analysed separately and the results are averaged except for the standard error term. A variation on imputation for categorical variables (or continuous variables that are categorised, e.g., into equal thirds or fifths) is to create a "missing category". This avoids having to drop records or guess the real value. The implication is that there's something special about this group, that is, that their outcomes are different from people in the other categories. This won't work so well if in fact they're just some mix of the other categories. Nevertheless, this approach avoids losing statistical power incurred by dropping records and has the potential to encourage the submission of better quality data.

There have been some empirical comparisons of the effect on performance estimates of either imputation or dropping cases. In a simulation study, Glance et al. (2009) assessed how different methods for handling missing data may influence which hospitals are designated as quality outliers for trauma mortality. Starting with real data from the National Trauma Databank, patients were assigned missing data for the motor component of the Glasgow coma scale (GCS) conditional on their observed clinical risk factors. Three approaches were compared with the results from the true data: complete-case analysis, multiple imputation and excluding the risk factor with missing data. With 10% of the data missing, the level of agreement between multiple imputation and the true data (intraclass correlation coefficient [ICC] = 0.99 and kappa = 0.87) was better than the level of agreement between complete-case analysis and the true data (ICC = 0.93 and kappa = 0.62). Excluding the predictor (motor GCS) with missing data from the risk-adjustment model resulted in the least amount of agreement with quality assessment based on the true data (ICC = 0.88 and kappa = 0.46). Regarding whether a hospital was classified as a high or low outlier based on whether its observed-to-expected mortality ratio significantly differed from 1, complete-case analysis misclassified nearly 15% of the hospitals, whereas imputation misclassified only 6% of them. The patterns were the same with 5% missing data. The authors came out in favour of multiple imputation over exclusion. Earlier work not related to hospital report cards also found that, when the data are MAR, complete-case analysis will produce biased risk-adjusted outcome measures (Donders et al. 2006). There's no substitute for top-quality data, but some deficits can be ameliorated to some extent.

We should remember that missing values also occur in fields that appear complete. This can happen if data entry software uses some default value if you don't enter anything, for example, "no" to questions such as, "Does the patient smoke?" Among the most recognised situations in health services research is with secondary diagnosis codes in hospital administrative databases. Various factors such as coding rules, incomplete documentation and variations in coder practice and workload conspire to cause under-recording of comorbidities at many institutions. Numerous studies have assessed the extent of this and how it impacts on the validity of, for example, Agency for Healthcare Research and Quality's patient safety indicators (Sundararajan et al. 2015 and references therein). To summarise what was covered in Chapter 5, the specificity is generally high or very high, though the sensitivity is often moderate. This is why researchers have incorporated information from previous health service contacts, which has its own issues (see the next section).

Extreme values can also cause a problem. Implausible values may need to be treated as missing. Extreme but plausible values such as very high costs will distort predictions for patients with non-extreme values. Options for handling them include dropping the cases, winsorising (keeping the cases but replacing the extreme values with something unusual but less extreme, e.g., replace very high values with the 95th percentile) and transforming, for example, taking the natural logarithm (this can give a "nice" normal distribution but back-transforming onto the original scale may either be impossible or, with exponentiating logged values, change the interpretation of the coefficients). If the extreme values are plausible, then the best option depends on the goal of the analysis. With risk adjustment, if the extreme values affect a risk factor, it may be possible to accommodate them in the model just as they are. The usual residual plots and checks for influential and outlying points should be made. Winsorisation is a good option if the exact size of the value isn't important, merely that it's "big" (or "small").

6.7 Timing of the Risk Factor Measurement

With some databases and variables, the patient can have multiple measurements over time. With physiological variables and lab results, the worst value might be the most predictive but could also reflect poor care. The first value is probably the easiest to spot in the database and will better reflect how the patient was when they presented to the healthcare service. In any case, even if there is only one measured value, the timing at which that measurement was made relative to the outcome needs to be considered. That first value might be too remote from the outcome, though the first value after admission may be the best for use with hospital outcomes. An exception to that would be for ITU outcomes, when risk scores such as Acute Physiology and Chronic Health Evaluation (APACHE II) are applied within 24 hours of admission to the ITU.

It's common to use previous health service contacts as risk factors. For hospital outcomes, one way is to augment the secondary diagnosis fields with information from previous admissions or physician claims to boost the comorbidity prevalences. A "lookback" period of a year is used by the CMS for this purpose, though other lookback lengths have been tried (Preen et al. 2006). The evidence on what is the best length is mixed, so a year seems reasonable. One drawback of this is that it won't give any information for people with comorbidities who simply haven't used the health service, which will add an unknown level of bias. A second one is that, as mentioned earlier, databases generally give little idea of how severe, active or controlled a disease is. In the United Kingdom as in some other countries, coders are meant to record only those comorbidities that are "relevant" to the primary diagnosis on admission. For example, diabetes that was recorded in an admission a year ago but not in the index one may now be well controlled and have little impact on the chance of the outcome of interest occurring. In theory, inclusion of information from prior contacts could actually dilute the effect for that comorbidity on the outcome of interest and worsen the performance of the risk-adjustment model, though this has not been our experience.

Also regarding hospital data, the choice of patient record can make a difference to the number of records included in the analysis and to the unit's apparent performance. This is a particular issue with UK hospital databases such as England's Hospital Episodes Statistics (HES), where an admission can be represented by more than one record (called a consultant episode). This is because another record – and therefore another row in the database – is generated in HES whenever the patient is transferred to the care of another senior physician (consultant). The first record could relate to the short stay in an acute medical unit and the second to the inpatient ward. Each record has a set of data fields for the diagnoses and procedures that relate to that part of the hospital stay and can therefore differ from record to record within the same admission. We generally only want to count each admission once, so we need to choose the information of most interest from the two (or more) records. Suppose we are interested in counting AMIs. The first record might only say "chest pain" (ICD10 R07) before the diagnosis of AMI is confirmed and put down in the second record (ICD10 I21). If we take the first record of the admission, we'll miss this AMI. It may be better in general to ignore symptom codes (ICD10 R chapter) in the first record if the admission has more than one consultant episode (Bottle et al. 2010). The assumption is that the patient came in with the AMI rather than with, say, non-cardiac pain that was only followed by an AMI as a result of the staff's actions, that is, it was a complication of treatment or even that it just happened naturally after admission. With reliable present on admission information, which HES lacks, it's easier to make the distinction.

The chance of a patient going on to have the outcome of interest is of course influenced by the sum of their experiences (genes, lifestyle, diseases and treatments) right up to the end of the follow-up period for the outcome. Their odds of dying within 30 days of surgery are affected by their pre-op, intra-op and post-op experiences, including the quality of the care they receive. Our task is to *separate (and not adjust for) markers of that quality that relate to the unit that we are evaluating during the episode of care that we are including in our analysis*. It's worth unpicking that piece by piece.

Take the example of mortality within 30 days of a coronary artery bypass graft (CABG). Let's say that it's a revision procedure which is needed because of poor surgical technique by the same surgeon during the first CABG. Suppose we take into account the fact that this is a revision procedure, which has worse outcomes on average than first-time ones. Let's also ignore the fact that it was necessary only because of this surgeon's poor quality of care the first time round. We are now only considering this revision procedure, not what led to it. We should not adjust for events occurring during or after the revision such as complications that are related (at least in part) to the quality of care concerning this procedure. It would, however, be appropriate to adjust for intraoperative factors such as surgical complexity (e.g., the fact it's a revision and whether the surgeon planned to repair the mitral valve as well) but not for unexpected activity such as administering blood following a haemorrhage. What about pre-operative factors – are they all fair game for the adjustment model? Not necessarily. Whether or not antibiotic and venous thromboembolism (VTE) prophylaxis was given will influence the odds of mortality, but this is under the control of the surgeon under evaluation and so should not be included. What about drugs prescribed in outpatients or elsewhere before admission for the CABG? Such information could tell us something about the number and possibly also the severity of the patient's diseases, but it also tells us something about the quality of care at that time. It's convenient to divide information like this into that during and that before the index admission, even if the drugs were prescribed by the same surgeon, for instance, in an outpatient clinic. Pre-admission factors can be included if they are considered or empirically found to influence the outcome, whereas intra-admission ones need more careful consideration. To summarise, previous healthcare experience is worth considering for risk adjustment but only after thinking through the possible influences and biases.

Before we describe how to go about building the model, Table 6.1 gives a summary of the issues associated with the common risk factors given earlier.

TABLE 6.1

Issues Associated with Common Patient Factors in Risk Models

Factor	Issue	Comments
Age	For elective and especially paediatric surgery, age may be under the unit's control	When age affects outcome of surgery, adjusting will hide some quality of care effect.
Age	Fit as continuous or categorical	It depends on relation with outcome and whether model coefficients need to be interpreted.
Socioeconomic status or deprivation	Proxy for lifestyle and health status	Usually area-level measure, so may not describe individual patient. Adjusting will hide some quality of care effect if inverse care law applies: stratification and sensitivity analysis are useful.
Comorbidity	Choice of weight	Published weights are easiest but are probably suboptimal for your data and outcomes.
Comorbidity	Measurement period	Index event (e.g., admission) or lookback period? Studies differ and there can be biases due to access to care. Can be difficult to distinguish between comorbidities and complications in administrative data.
Comorbidity	Data source	Patients can give different answers from casenotes. Admin data are famous for under-recording: this may in part relate to record-keeping, which in turn relates to quality of care.
Disease severity	Data source	Usually hard to get from administrative data. Patients' ratings tend to be more severe than doctors'; patient survey data quality is influenced by non-response bias, cultural differences, etc.
Physiological values	Measurement error	BP subject to digit preference or gaming "Upcoding" can affect comorbidities, palliative care use, etc., as well if it triggers higher DRG/Healthcare Resource Group or P4P payment.
Physiological values	Timing of measurement	The worst value is most predictive of outcome but can reflect poor care. The first value is probably easiest to spot in a database. Consider whether timing is too remote from outcome.
Previous health service contacts	Whether to include them	Previous admission is the strongest predictor of future admission. As with lookback periods for comorbidities, this can add unknown bias due to differential access to services. It also requires more data years.

6.8 Building the Model

Having processed the data to deal with missing and extreme values and arrived at a list of candidate variables for risk adjustment, we have the following tasks:

1. Choose the final set of variables, including whether any interactions should be tried.
2. Decide on the statistical method for modelling.
3. Assess the fit of the model.
4. Output observed and model-predicted outcomes for computing the risk-adjusted outcome measure of choice.

6.8.1 Choosing the Final Set of Variables from the Initial Set of Candidates

As mentioned in Section 6.5, some can be ruled out because of known problems with recording and accuracy. Now we can also rule some out if they occur too rarely in our data set, though we may still want to try to include them in order to make the results acceptable to users. A popular rule of thumb says that we need 30 cases per predictor if the outcome is continuous (and cases should outnumber predictors by 10 to 1) and 10 cases with the outcome per predictor if it's binary (Shroyer et al. 1998) though a ratio of 20 is safer. A mixture of clinical judgment and empirical modelling is probably best, at least as a sense check to establish face validity. For reasons of acceptability or a priori knowledge, we may have a few that we want to include even if they don't turn out to be statistically associated with the outcome. For the rest, a common approach is to look at bivariate (crude) associations, for example, chi-squared tests between a comorbidity variable and mortality, and bin those where $p > 0.1$ or some other arbitrary threshold. Some developers prefer 0.1 or even 0.2 as the cutoff in order not to exclude an important variable due to having only a modest sample size.

At this point, developers commonly divide the data set into two or three portions if it's large enough: training, testing and validation subsets. With three portions, the training set is used to create a number of different models, which are each applied to the test set and compared using metrics such as the R^2 that are described later. The method that performs best is then applied to the validation subset to check that the model can generalise to previously unseen data. The validation subset can be "internal", that is, part of the same chunk you began with, or "external", such as from another unit or country. Risk-prediction models are frequently developed using one hospital's local data. If the model is to be used elsewhere, it must be tested on another hospital's data. With risk adjustment, however, this is not necessary unless that model is to be exported to another region or country where the database and patient mix are different. For a national monitoring system that has national data and requires its own custom risk model, internal validation is sufficient. With internal validation, if the data are only split into two or three chunks, then the model performance can depend heavily on which data points end up in the training set and which end up in the test set. The same problem happens if, instead of a single (internal) test set, a single external test set is used. If the data sets are large, however, this shouldn't cause a problem.

With only a small sample size that's too small to divide into such portions, then developers can use workarounds such as cross-validation and bootstrapping; these can be used with large databases too, though computational time could be an issue. Cross-validation is the simplest technique and in many ways the best accepted. These methods all involve generating subsets of the original data for internal validation. Not ideal, but much better than nothing. For example, in leave-one-out cross-validation, we first split our data into two chunks: one with a single observation and the other with all the remaining observations. We fit our model to the second chunk and use it to estimate the outcome for the observation that's been left out. Then compute the error for this observation, which is just the difference between the predicted outcome and the actual outcome. Then repeat the process, leaving out a different observation, and continue until all observations have been left out of the modelling once. The mean squared error (MSE) is the average of all the individual errors from this process. This is a common measure of the accuracy of a linear model. For models where our outcome measure is binary, for example, we want to classify patients as either alive or dead, the misclassification error rate could be used instead. (For linear models, there's a shortcut using the "hat matrix".) As a variant on this process, we can leave out k observations each time rather than just one (called leave-k-out cross-validation). Another common variant is to randomly split the data set into k subsets (called k-fold cross-validation), where k is often chosen to be 10. In this case, the process should be repeated a number of times, for example, at least 50 times for 10-fold cross-validation. Yet another is to first stratify the data set on the outcome variable before splitting (stratified cross-validation), and still other variants exist. The resulting mean squared error (MSE) from these processes is a good way of choosing the best-fitting model (pick the one with the smallest MSE); although a lot of models (algorithms) are being tested, this approach can lead to problems. Leave-k-out cross-validation won't work very well if there are multiple patients with exactly the same values for their outcome and predictors. This is because there is effective overlap (lack of independence) between the patients used in both the training and testing chunks. Leave-one-out cross-validation also has well-known problems of underestimating the true error and retaining noise variables. Much theoretical and applied work has been done on cross-validation, but there is still much to learn.

Armed with our list of candidate variables, they can then be entered using various "stepwise" procedures. Stepwise can be forwards, backwards or both, with the latter often being called simply "stepwise". Variables can be entered one at a time, beginning with the one with the largest expected contribution ("forward selection") or in a kind of enter-test-drop dance ("stepwise selection"). Both these model selection methods are common despite being fraught with well-known difficulties, such as resulting in models that can be hard to reproduce and that can yield biased coefficients. They should be avoided where possible, particularly if you have fewer than 1000 records (Harrell et al. 1984). The third stepwise approach is to enter all the candidates and drop the non-significant ones ("backward elimination"), which is likely to be superior to the other two (Mantel 1970), especially when some variables are correlated with each other (a problem known as multicollinearity). We've found it to work well when there are a large number of models to be fitted (Jen et al. 2011), as in our national monitoring system (Bottle and Aylin 2008a). However, it's common for models resulting from any of these methods to retain noise variables (Austin and Tu 2004). This is more of a problem if the goal of the modelling is explaining or predicting risk than for adjusting risk, though there could be a cost to face validity.

Interactions between pairs or even larger numbers of variables can add extra explanatory power to the model. The trouble is that often they don't and there are potentially a lot to try. Clinical judgment can really help here to produce a short list. Empirical testing can then decide which to retain.

There is a cost to all this testing that can easily be overlooked: type I error, or the chance of seeing an association between two variables that doesn't actually exist and hence retaining "noise" variables. This is common (Austin and Tu 2004). Such spurious associations can spring up due to random or non-random fluctuations in the data. It's a particular problem with forward selection and stepwise procedures, which is another reason to avoid them.

Two other more recent approaches to reducing the problems of having multiple correlated variables and having too many candidate variables are shrinkage techniques and hybrid techniques. The former are particularly good for multicollinearity and include ridge regression and lasso regression (Tibshirani 1997; Steyerberg et al. 2000). They only include a portion of each of the variables, corresponding to the variables' importance. Hybrid techniques combine a way to create new variables that aren't correlated with each other with regression as usual. An example is principal components regression. This creates the new set of truly independent input variables using principal component analysis and then uses these in a logistic regression model. These and other methods are briefly described elsewhere in the context of their usefulness for National Surgical Quality Improvement Program (NSQIP) (Fabri 2014).

Another approach, which has its origins in the 1960s, is model averaging. In this, all possible models, including those with interactions between variables, are considered, and their coefficients are averaged across them.

This is usually done in a Bayesian framework despite some associated difficulties (Hoeting et al. 1999) as Bayesian model averaging (BMA). There's a BMA package in R to implement it. The great statistician George Box said that all models are wrong, but some are useful. The concept of BMA is therefore appealing. Non-BMA approaches rather give the impression that they have definitely yielded the best one, which may not in fact be true. Simulations show that in most cases Bayesian model averaging selects the correct model and outperforms stepwise approaches (Wang et al. 2004; Genell et al. 2010). For an explanation of the Bayesian interpretation of probability and therefore what "Bayesian" means, see Appendix A.

6.8.2 Decide How Each Variable Should Be Entered into the Model

Continuous variables pose the question of how they should be modelled. At the very least, you should plot the relation between each one and the outcome. Expert opinion can also inform whether, for instance, a linear relation is likely to be valid or some kind of U-shaped relation needs to be tried. Particularly if the coefficients don't need to be interpreted, polynomials or cubic splines can be applied (Royston et al. 2009). If the relation is non-linear and there is a desire to interpret the coefficients, then multiple categories can be used, for example, five-year age bands. This will probably work nearly as well as something much more complicated but does come at the cost of needing more data to accommodate the extra coefficients. Sparsely populated age groups may need to be combined. Another drawback is that it's harder to add interactions between age and other factors compared with having just a single term in the model for age as in the linear relation case.

6.8.3 Decide on the Statistical Method for Modelling (*)

Regression methods are the most common by a country mile. They are relatively easy to understand, can be run in all the standard software packages and produce coefficients that tell us the relation between each predictor and the outcome. If using regression, we need to decide whether to account for clustering of patients within unit or any of the other kinds of clustering discussed in Chapter 3 on deciding the unit of analysis. That chapter covered marginal or population-averaged models using generalised estimating equations, but performance summaries such as physician or hospital report cards generally use hierarchical modelling if they do adjust for clustering, so we will cover that in this chapter. For example, as will be described in more detail later, the CMS and NSQIP use hierarchical logistic regression for risk-adjusting mortality and other outcome rates by hospital. We need to consider the eventual output when deciding which statistical method to use for risk adjustment. There are two parts to this output. One is the risk-adjusted outcome measure, such as the proportion of patients with post-operative complications by surgeon, and the other is how that measure will be reported, for instance, in performance bands of above expected, as expected and below expected. This chapter will consider the first, with the second to be covered in Chapter 9.

In principle, dealing with clustering is the "correct" thing to do, but unless you want to use these models' "shrunken estimates" (more on that later), it's not always necessary in practice. Simulation studies have shown that ignoring the clustering leads to standard errors that are too small and an inflated type I error rate, meaning that more variables appear statistically significant than they actually are (Tate and Wongbundhit 1983). In risk adjustment, however, we're not interested in the standard errors of any of the risk factors, just the predicted value for each patient (the risk factor coefficients and hence the overall predicted value are not biased). The exception to this would be if we decided to use some variable-selection procedures such as stepwise selection that use the standard errors but, as we discussed earlier, these are fraught with problems and should be avoided wherever possible. In our experience, the degree of clustering is often small with hospital but not primary care data. If this is true for your particular situation, then there seems little benefit in spending the extra effort specifying the model, running it and getting it to converge (hierarchical models require more computer memory and CPU time). It also requires sufficient numbers of observations to estimate the clustering, though simulations have shown that as few as five observations per group, for example, five patients per clinic, will yield valid and reliable estimates of all parameters and their standard errors (Clarke 2008).

Despite those difficulties, sometimes we need to use multilevel models and sometimes we want to. If there is any clustering, no matter how small, it can always be considered "correct" to adjust for it though perhaps not necessary. If the degree of clustering is not "small", the case for adjusting for it can become compelling. So what counts as "small"? How much clustering is enough to worry about? Another rule of thumb for linear models suggests that an ICC of >5% suggests that we should deal with the clustering (Degenholtz and Bhatnagar 2009). The ICC describes how strongly subgroups in the same group resemble each other, for example, how similar the outcomes in patients at the same hospital are. It tells you the proportion of the total variance in the outcome measure that is accounted for by the clustering. The maths is much easier for linear models, though various methods exist for non-linear ones such as are needed for binary outcomes (Fitzmaurice et al. 2004). An ICC of zero means that we definitely don't *need* to worry about clustering, though we still might want to fit a model that can account for it. One advantage of multilevel models is that they assume that the distribution of *unmeasured* patient factors can vary between hospitals. This is a good idea in theory, given that no data set contains all the factors we'd like to adjust for.

Although regression methods are standard, there are a large number of more sophisticated "machine learning" techniques. There is both a growing list of these and a growing literature on comparisons between them, including in the past 20 years or so in healthcare (they were generally developed in other spheres).

Here is a list of only what seem to be the most common in healthcare: artificial neural networks (ANNs), support vector machines (SVMs), decision trees (and extensions such as random forests), k-nearest neighbours and genetic algorithms. They have long been employed in medicine with the goal of classification, for instance, in cancer diagnosis, but also for prediction and prognosis from around the 1990s onwards. They need either specialist software or R modules in order to implement up-to-date versions.

As with any method, they each have pros and cons. For example, compared with regression, ANNs are more flexible, can handle more variables and can model complex non-linear relations with fewer assumptions, but there are no set methods for constructing the architecture of the network (the most common type of ANN is the feed-forward back-propagation multiperceptron). Rather than produce coefficients such as odds ratios, they produce weights, which are less interpretable. Unlike with regression, machine learning methods can yield different answers depending on factors such as starting values and their hyperparameters (extra things to tune, which regression doesn't have to worry about) (Matheny et al. 2007). Overfitting is one of the main risks with these methods. This is when the machine learns which data features are associated with the outcome so well that, when confronted with another data set, it predicts poorly when it fails to find the exact same data features in the new set. This is why it's standard practice to split your data into two or even three portions as part of the modelling. We compared ANNs, SVMs, random forests and logistic regression for mortality and readmission using comorbidity flags and a few other fields taken from administrative data and found the results fairly disappointing (Bottle et al. 2014). A 2002 review of 72 papers comparing ANNs with logistic regression found that many comparisons were equivocal, but the more flexible networks did do better (Dreiseitl and Ohno-Machado 2002); this review is also useful as an account of both methods in detail with brief descriptions of some other popular machine learning approaches. Since then, other researchers have continued to find the methods promising in risk prediction (Eftekar et al. 2005). Most studies focus on prediction of risk rather than our goal of adjusting for risk, though the distinction is not important for our purpose here of describing the range of available approaches.

One recent study that is specific to risk adjustment concerned in-hospital mortality in neonatal intensive care units (Liu et al. 2015). This compared random forests, Bayesian additive regression trees (BART) and logistic regression and found that the first two did better on both accuracy of estimated risk (Brier's score, false positives and false negatives) and discrimination (see later section), although in this particular data set all methods performed very well. The authors then went the necessary extra step of showing what difference the set of risk estimates made to hospital performance as measured by the ratio of observed to predicted deaths. They gave the percentage of hospitals that changed from one quintile to another when different risk-adjustment methods were applied. There was considerable movement between quintiles even though dramatic changes such as from top to bottom were rare. For example, in the worst-performing bottom fifth of hospitals, 70% of those classified as such by logistic regression were also thus classified by BART, 22% were put into the next fifth up by BART and none was put into the top fifth. It only needs a small change in the measure's value to move band or rank if the unit is near the boundary between bands or, as is often the case, there are a lot of units with similar values. We would therefore expect some movement with this method of comparing units, particularly between neighbouring quintiles, but it's likely that there will be a few unlucky hospitals whose estimated performance depends on the risk adjustment, whatever method is used to classify their performance (see Chapter 9 on defining outliers).

For a non-technical review of the common types of machine learning methods, their benefits, assumptions and limitations, see, for instance, Cruz and Wishart (2006). They review the machine learning methods used in cancer prediction showing the types of cancer, clinical endpoints, choice of algorithm, performance (as percent improvement over the comparator) and type of training data, and conclude that, "it is clear that machine learning methods generally improve the performance or predictive accuracy of most prognoses, especially when compared with conventional statistical or expert-based systems".

Machine learning methods will perform better with richer data where there are fewer ties (patients with exactly the same profiles but different outcomes). We believe that to really extract the most explanatory power from your database, unless it has only a few possible predictors, some kind of machine learning technique will be needed. However, there is a shortage of people who understand them, and, due to their (slightly unfair) "black box" label, they are therefore unlikely to take hold in P4P schemes or publicly reported measures any time soon. From here on in this chapter, we'll assume that regression has been used for the risk adjustment.

6.8.4 Assess the Fit of the Model (*)

This tends to be done iteratively as part of model fitting as well as at the end. We need some measures of how well the model fits the data. Many of them can also help choose between candidate models. The most commonly used methods to choose among classification methods are mean squared error, adjusted R^2, Mallory's Cp, and the Akaike information criterion (AIC). Linear models can be assessed using the R^2, residuals, outliers and influential points. (Note that "outliers" in this case is different from units that are outliers in terms of their performance, as in Chapter 9 later.) The most common measures for assessing a logistic regression model are the c statistic (area under the receiver operating characteristic [ROC] curve) and the Hosmer–Lemeshow statistic. We can't discuss all these in detail, so we'll focus on the three that probably cause the most debate and misunderstanding: adjusted R^2, c statistic and the Hosmer–Lemeshow statistic.

6.8.4.1 Adjusted R^2

In regression in general, R^2 measures quantify the extent to which the outcome can be predicted by a given regression model and covariates. With linear regression, it has a nice interpretation: it estimates the proportion of the outcome accounted for by the model and takes on values between 0 and 1 (or 0% and 100%). Values can be high. With non-linear regression such as for binary outcomes, however, the maths doesn't work so neatly (e.g., the maximum possible value is no longer 100%). Different R^2-type measures have therefore been proposed. Some can be adjusted for the number of parameters, as well-behaved R^2 measures increase with each new variable that's added unless its parameter estimate is zero. Mittlböck and Schemper (1996) reviewed 12 popular versions and recommended three of them (the default in the statistical package SAS was not among the three). Hu et al. (2006) then investigated these three and preferred the sums of squares and the squared Pearson correlation measures. These are types of variance measures, that is, they try to copy the linear model R^2 in order to estimate the proportion of the variance in the outcome that the model accounts for. A second type is based on something called the entropy or information gain: these are called pseudo-R^2s and are popular in the social sciences. Hu et al. (2006) point out that only the variance-type ones have properties akin to those in the linear model case. Shtatland et al. (2002) recommend the McFadden (or "deviance") pseudo-R^2 and provide an AIC-type adjustment for the number of model parameters. Whichever R^2 version is chosen, it's better not to use it as a goodness-of-fit measure – other tools such as calibration are more suitable for that.

R^2 values in logistic regression are habitually low, say, <0.3. This can disappoint and frustrate those who are used to seeing much higher values for linear models. The outcome variable can be only 0 or 1, whereas the model's predicted probabilities will almost always be somewhere in-between. Mittlböck and Heinzl (2001) have shown that, when the binary outcome is rare enough for the Poisson to be a good approximation to the binomial, the R^2 will be much higher for the Poisson than for the logistic regression even if the same covariates are used in the same way. This is because it's easier to predict the total number of deaths (the task of a Poisson model) than whether an individual patient will die (the task of a logistic model). The R^2 estimate for the Poisson regression quantifies how well the model predicts the event rates, not how well it predicts the outcomes of individual patients.

6.8.4.2 Area under the Receiver Operating Characteristic Curve: c Statistic

Discrimination is a measure of how well the model can separate those who do and do not have the disease or outcome of interest. It's most often measured by the area under the receiver operating characteristic (ROC) curve, or c statistic. Using the language of a diagnostic test, this curve is a plot of the sensitivity of the test to detect the disease against the false-positive rate of the test.

Perfect discrimination corresponds to a c statistic of 1, when the scores for all the cases are higher than those for all the non-cases and there's no overlap. A value of 0.5 indicates no discrimination and a worthless diagnostic test. In risk-prediction/risk-adjustment papers, the c statistic tends to take centre stage in the abstract and results sections. Sometimes, it's the only measure of model performance that's given. In classification problems, when we aren't interested in the size of the predicted risk, this can be appropriate. For risk adjustment, however, it's rarely the case.

The c statistic is not a function of the predicted probabilities themselves, which is why we also need to assess calibration. It also means that it's a poor tool for choosing between risk models, though unfortunately it's sometimes employed for that purpose. A risk factor with an odds ratio as high as 3 might not noticeably change the c statistic (Pepe et al. 2004). Another fact that deserves to be more widely appreciated is that, unless the real risk of the outcome being predicted is either 0 or 1, there is an inherent trade-off between discrimination and calibration. With a perfectly calibrated model, the c statistic has a maximum value of less than 1, potentially much less than 1, depending on the distribution of risk in the population. The real risk of the outcome could be 0 or 1 in the diagnostic setting, when the person either does or does not have the disease, but not in our setting of risk prediction and adjustment. For a very readable discussion of the issues around using the c statistic, see Cook (2007).

6.8.4.3 The Hosmer–Lemeshow Statistic for Calibration

We want the model to fit the data well overall, but it's also important for risk prediction/adjustment to get a good fit across the distribution of risk. The Hosmer–Lemeshow test typically divides the data set into 10 (risk deciles), with an equal number of patients in each decile, which means that there are a very unequal number of outcomes by decile. The observed number is compared against the predicted number in each decile and a chi-squared statistic calculated, the resulting p value being obtained using eight degrees of freedom. P values above the conventional cutoff of 0.05 suggest no important deviation of the predicted from the observed, that is, good calibration. The lower the p value the worse the calibration. It's a convenient summary and much used.

The creators themselves recognise two limitations: poor power to detect miscalibration with small samples, such that important differences are missed, and hypersensitivity with large samples, such that unimportant differences are emphasised (Hosmer and Lemeshow 1995). With large samples, as is the norm with administrative data in particular, the plot of observed versus predicted by decile is probably more useful than the p value. Even more useful, but not as easy to get, is an estimate of how many extra or fewer outcomes the miscalibration would assign to each unit being monitored. For example, suppose we are using national data to estimate hospital death rates.

One could divide a hospital's set of patients into the national deciles and calculate the number of predicted deaths in each decile. Then multiply these predicted by the national observed-to-predicted ratio for the decile and subtract the predicted. With perfect calibration, the result will be zero for all deciles. A common concern, particularly with public reporting of surgeon-specific outcome rates, is that the model underestimates mortality for high-risk patients. The suggested calculation would yield a negative number in this situation. Hospitals (or surgeons) operating on a lot of high-risk patients would see a larger negative number than those treating low-risk cases. Their predicted number of deaths would be too small and their adjusted mortality rate too high. This kind of calculation tells us more about whether the risk model is fit for our purpose than a low p value from the Hosmer–Lemeshow test.

6.8.5 Which Is More Important, Discrimination or Calibration?

As usual, it depends. If the actual value of the predicted value isn't important, then calibration isn't either. For comparisons of healthcare units, however, calibration is considered more important than discrimination (Ash and Shwartz 1999). A high c statistic does not necessarily make a good hospital report card, at least from a technical point of view. Simulations of the relationship between the c statistic of a logistic regression model used for risk adjustment and the accuracy of hospital report cards have shown it to be modest (Austin and Reeves 2013). They also found that increasing patient volume was associated with increased accuracy and that this relation was stronger than that with the model's discrimination. This doesn't mean that we shouldn't aim for the best c statistic that we can extract from our data, but that it should not trump all other considerations. In terms of calibration, we should aim for good overall calibration, but it's perhaps more useful to see what it means for a typical unit, as suggested above and elsewhere (Bottle et al. 2013b). Summary statistical performance measures like the ones we've just discussed don't tell us how well risk-adjustment models work for different subgroups of patients. Therefore, if we're choosing among risk-adjustment methods we shouldn't just pick the one with the highest R^2 or c statistic but should also consider clinical validity. The ultimate test of the method is whether it helps reveal problems with quality of care rather than noise.

6.8.6 What Can Be Done If the Model Fit or Performance Is Unacceptable?

If it's a problem of model fit, such as lots of large residuals, then try a different specification, for example, the probit link instead of the logit link. If an external risk score is used and results in suboptimal calibration, then this can beZcorrected, for instance, by using a "fudge factor". Recalibration or even reformulation of a model is periodically required due to temporal changes (e.g., because of steady improvements in the quality of care or abrupt changes in disease definition and/or diagnosis) or the emergence of a new risk factor (which could simply be because of an additional field to the database). Calibration tends to deteriorate over time, an example being the Euroscore in cardiothoracic surgery (Hickey et al. 2013a). One simple option that will avoid abrupt changes in the estimated effects of a variable in the model is to use a rolling time window. For instance, if the database is updated monthly and 24 months of data are normally used to construct the risk model, then the most recent 24 months can be used. If the data infrastructure allows it, dynamic risk adjustment is also possible. McCormick et al. (2014) describe how to start from the usual model fitted to the training data portion and then use a Bayesian paradigm to update the model coefficient estimates for each subsequent time point. This uses the approximated posterior distribution at time t as the prior distribution for time t + 1 in the prediction equation. The approach was compared with the use of static or periodically updated models in a study of post-operative mortality (Hickey et al. 2013b). The extra complexity of the dynamic models, both in terms of fitting and describing their performance (which is itself dynamic), seems to counteract any gains in model performance.

If the model's predictive power is low, for example, with low c statistic and/or R^2 (but note the foregoing limitations of these measures), then we can try running sensitivity analyses for the effect of unmeasured confounders. This is useful if we know that our database lacks some key risk factor *and* that that factor varies considerably by unit. Model performance may be poor due to data noise such as measurement error, which is likely if we have to rely on proxies or if there have been a lot of data entry errors – or it can be poor due to real variations in quality of care. Alas, we can't tell the two apart from the data alone. A low c statistic could be due to wide variations in quality of care rather than (or as well as) important missing covariates. As we have seen with socioeconomic deprivation and the inverse care law, some covariates that predict the outcome may also relate to quality of care and should potentially be left out of the model, which will impair the model's apparent performance.

6.8.7 Convert Regression Coefficients into a Risk Score If Desired

Whether the outcome variable is continuous, binary or a count, the set of coefficients for age, sex, and so on, from the final risk model can easily be converted into just one: the risk score. Each coefficient is converted into a number of "points" (rounded as integers) based on its importance (size) relative to the model's least important variable, and the points are then totted up for each patient. For details, see Sullivan et al. (2004). While this incurs some rounding error, one can argue that using the precise estimate of the coefficient is being overoptimistic about the model's reflection of reality. Using a risk score has several advantages. It can be used to compare the casemix profile of units more easily than a host of categorical variables (though you might, for instance, want to compare operation complexity profiles by unit). This is particularly important when deciding whether units are similarly enough to allow a fair comparison. If a unit sees very different types of patients from the national average, then comparing that unit with the national average is inappropriate. For instance, no amount of adjustment would allow comparison of a paediatric hospital with an acute general hospital. Second, as we shall see later, it can be useful to rescale the score to have an average of zero. Third, as mentioned earlier regarding applying a "fudge factor", periodic recalibration is easier when the model contains just an intercept and a slope parameter, as is the case when all the risk factors are combined into a single variable, the risk score.

We now have our risk-adjustment model. The next chapter considers the various ways to extract our unit-level risk-adjusted outcome measures.

7

Output the Observed and Model-Predicted Outcomes (*)

Aims of This Chapter

- To summarise the pros and cons of using O/E or O minus E
- To describe the various ways of producing risk-adjusted outcome measures from the models in the previous chapter
- To use the calculation of standardised mortality ratios as an example

This is probably the most technical chapter in the book and contains the few formulae that we've felt necessary to include. As elsewhere, the most statistical sections are marked with (*), as is the whole chapter because of this. We go through how to produce a risk-adjusted measure for each healthcare unit, which compares the unit with the benchmark. This might be post-operative mortality or unplanned admissions by general practitioner. For the purposes of illustration throughout this chapter, we'll consider a binary outcome measure, specifically mortality, unless stated otherwise. Outcomes that can be assumed to have a roughly normal distribution, such as blood pressure, weight or patient experience scores, are often turned into binary measures such as "high" versus "normal". If they aren't, then things are a bit simpler, though many of the issues that we'll discuss still apply.

This chapter covers various methods for producing summaries of the observed (O) and expected (E). Section 9.7 details the research on how much difference the choice of method makes on whether a unit is labelled as good, bad or indifferent. As the short answer is "sometimes a lot", it's important to know the options.

7.1 Ratios versus Differences

When comparing the observed total with the expected total that the model predicts (hereafter called the "expected"), there are two distinct ways: ratio and difference. We can take the O/E ratio, which is often called a relative risk (RR) with binary outcomes or a standardised mortality ratio (SMR) if the outcome is death. Alternatively, we can take O minus E. The former is much more common in performance monitoring and often the highest profile, which is why we'll focus on it in this chapter. The latter is used more in epidemiology, for instance, when assessing the impact of risk factors on disease rates. Although ratios are more common, it's not clear which is best for our purposes. Consider this example. Of each pair, which unit performs poorest?

1. Unit A has a complication rate of 2%, double the rate of unit B with 1%
2. Unit C has a complication rate of 20%, a quarter higher than D with 16%

In the first example, dividing the rate for A by the rate for B = 2%/1% for a relative risk of 2. In the second example, dividing the rate for C by the rate for D = 20%/16% for a relative risk of 1.25. Using relative risks makes it seem that the disparity between the two units is much greater in the first case.

However, the *difference* in rates is much more dramatic in the second: 20% − 16% = 4% compared with 2% − 1% = 1%. When applied to the population of such units, 4% represents many more patients having complications than 1%. Both the ratio and the difference are useful. Each should be given alongside their confidence interval (CI) to show how much random variation there is. Each has its limitations. In Section 9.3, we pick up on the dangers of the easy interpretation of the difference in rates as the number of "excess complications" or "excess deaths": there can be a lot of random errors associated with the estimate of (O minus E). The ratio also has drawbacks. O/E is not symmetrical around 1 as of course it can't go below zero, although taking the natural logarithm and plotting on a log scale gets round this: the sequence 0.25, 0.50, 1.00, 2.00 and 4.00 represents successive doubling, and the distances between them are evenly spaced when a log scale is used. A unit with a ratio of 0.50 will therefore appear to be just as relatively different from the average (1.00) as a unit with a ratio of 2.00 – half the average rather than double the average. This is not the case if you don't use a log scale. A second limitation of the ratio which we discuss in Section 7.2 is that it's sensitive to small changes in O when E is small. A common way around this is to ignore ratios with $E < 5$.

7.2 Deriving SMRs from Standardisation and Logistic Regression

Let's consider the mortality in a set of geographical areas such as regions of a country (our units), with risk adjustment performed by logistic regression. We might want to know which regions have significantly worse five-year survival following a given diagnosis (for simplicity, we'll ignore debates around the use of cause-specific death rates in cancer, for instance). Let's also assume that we have national data and that we want to compare each unit with the national average. The simplest unit-level output is the observed-to-expected ratio from a non-hierarchical model. This is a relative risk, as it's the ratio of the unit's death rate to the national death rate. It's often called the standardised mortality ratio (SMR) as it has a lot in common with indirect standardisation used in epidemiology that produces SMRs. With *indirect standardisation*, we need national death rates for each combination of age, sex and any other categorical variable for which we want to adjust. We also need the number of people by age, sex, and so on, in our unit of interest. Suppose we only want to adjust for age. With indirect standardisation, the calculation proceeds as follows:

1. For each age group, calculate the death rate in the country as a whole.
2. Multiply this rate by the number of people living in the unit in the same age group to get the expected number of deaths.
3. Sum the expected deaths across age groups in the unit.
4. Sum the total observed deaths in the unit and divide this by the total expected deaths to get the SMR.

SMRs are often multiplied by 100 and rounded to remove the decimal places. Units with SMRs of over 100 have higher death rates than expected. Although the whole country is the most common choice for the "standard population", that is, that against which each unit's rate is compared, there are other options that could be more relevant depending on which units are being compared. Note that the European Standard Population and the younger WHO World Standard Population can't be used with indirect standardisation because the observed counts don't exist for these artificial populations. This method is only really feasible for a small number of variables and can't handle continuous risk factors. We will use the term "SMR", though, to refer to *any* observed-to-expected ratio of binary measures, even if the expected number comes from a different method.

Now let's consider a logistic regression model for mortality. Again, suppose we're interested in comparing our country's regions in terms of their patients' five-year survival following some diagnosis. Standard software readily produces a predicted probability of death for each patient from this model by converting odds into probabilities. These predicted probabilities are summed to give a total number of expected deaths, E, for the unit (region in this example). We sum the total number of actual or observed deaths, O, for the unit, and get O/E as the SMR. As with indirect standardisation, it's often multiplied by 100 and rounded. It's interpreted in just the same way, so SMRs over 100 represent mortality that's higher than average. This method is also the most flexible in that it's easy to sum by different levels of aggregation such as surgeon, department, provider, day of the week and risk decile to allow further investigation if the portal for viewing the results is sufficiently flexible (Bottle and Aylin 2008a); another advantage is that it's computationally less intensive than other modelling methods. As with an SMR from indirect standardisation, this ratio is equivalent to the death rate at the unit divided by the death rate for the whole data set (i.e., whole country), adjusting for differences in the distribution of risk factors between the unit and the data set as a whole:

$$SMR_j = \frac{\Sigma_i O_{ij}}{\Sigma_i E_{ij}} * 100 \text{ for patient i and unit j}$$

As mentioned earlier, the ratio O/E can look suspiciously high when E is small. To use SMR-type measures, E should therefore be at least 1 and some people say at least 5, especially when displaying SMRs on funnel plots – see Chapters 9 and 11. The SMR can be multiplied by the national crude rate to give an adjusted death rate (Hosmer and Lemeshow 1995).

As with indirect standardisation, there's an assumption here that O/E is constant for all risk factors included in the adjustment. If the SMR is 200, then it implies that the unit has twice the death rate of the national average. It's assumed that this is true both for young and for old patients, for males and for females, for those with diabetes and those without. This is unlikely to be exactly true, and for some patient subgroups and units it may well be way out. How much of a bias this is will vary, but even those who are critical of this method give examples that show the bias to be generally pretty small (Julious et al. 2001). Whether an SMR is 130 or 135 is immaterial – unless of course it tips your institution over some adverse P4P threshold and costs you money in penalties. This assumption of constant rate ratios across risk subgroup such as age is most usually known as proportionality and is easy to check empirically.

Some commentators draw attention to a long-recognised theoretical problem known as Simpson's paradox. This occurs when the ratios for all the patient subgroups are actually constant but the distribution of the subgroups between units is markedly different, for example, the proportions of patients in each age group differs between the unit and the benchmark. The O/E ratios for all the subgroups are the same, but when you take the total O and divide by the total E to give the SMR, you no longer get the same value as each of the subgroup O/E ratios. Simulation shows that the effect can be considerable in theory (Marang-van de Mheen and Shojania 2014). Again, the degree to which this is a problem in practice is usually unknown (though it can and, one might argue, should be checked empirically) and it's unlikely to invalidate the measure by itself.

A natural way around this is to use another old method, *direct standardisation*, which has recently been proposed for risk adjustment (Nicholl et al. 2013). Instead of applying national death rates for each patient subgroup to each unit's population to get an expected number of deaths, we apply the unit's subgroup-specific rates to the national population to get an expected number. With direct standardisation, the calculation proceeds as follows:

1. For each age group, calculate the death rate in the unit.
2. Multiply this rate by the national total population in the same age group to get the expected death count for the national population.
3. Repeat and sum the expected deaths across age groups.
4. Divide by the total observed deaths in the country to get the SMR (in epidemiology, this tends to be called the comparative mortality figure).

Although this sidesteps the problems of indirect standardisation, it has two main drawbacks. First, small numbers in the unit's subgroups make for unreliable or "noisy" estimates when the unit's rates are multiplied by what will often be large national population counts for the patient subgroup. The second is that units can be judged harshly on patient subgroups that it rarely sees. Suppose there are just two risk groups, low and high, and that a unit has 1 high-risk patient who dies and 99 low-risk patients of whom 5 die. When these rates are then multiplied by the national population, its apparent death rate could be enormous and misleading, depending on the mix of the standard population. The method will work better with less extreme examples and decent numbers in each patient subgroup in each unit. As with other approaches, it shouldn't be judged harshly on the basis of one or two extreme, theoretical examples. Today, direct standardisation is rarely used to profile physicians or hospitals but is worth further evaluation. It's commonly used in epidemiology and public health.

To get our 95% confidence interval (CI) for our SMR, we invariably ignore the random error around E and assume that the national rates are based on numbers that are so large that their error is negligible. It might not in fact be negligible, but I'm not aware of any monitoring system that takes it into account. SMRs can be easily converted into adjusted death rates as mentioned earlier, and CIs are correct if we assume a binomial distribution for these rates. Analysts often assume a Poisson distribution if death is "rare" (which typically means <5%, but some use the Poisson for higher rates as well). For large samples, we can use the normal approximation and the square root of O in the CI calculation. For small or indeed any sized samples, Byar's approximation can be used (Breslow and Day 1987). This was done for CABG mortality in New York's pioneering monitoring system.

7.3 Other Fixed Effects Approaches to Generate an SMR

This is far from the only way to generate an SMR. DeLong et al. (1997) review eight ways to obtain a risk-adjusted death rate, in their case for CABG units. The eight are divided into those that use "external" sets risks derived from a published algorithm rather than the data at hand (cardiac surgery has had several predictive risk scores over the years) and those that derive the risks "internally", that is, from the hospitals' own data. They are further split according to whether modelling is performed and whether fixed or random effects are included in the models: we'll describe these terms shortly. If the risks are taken from an algorithm (published risk score), then no risk modelling is needed: the first method just uses the O and then the E from the algorithm, and the second makes a simple but slightly dodgy adjustment for the overall difference in death rates between the set of hospitals at hand and the estimated death rate for those hospitals used to create the algorithm. The third method improves on the second by deriving a new intercept and coefficient for the algorithm's risk score from the data at hand – the published risk score is *recalibrated* for the data. This is necessary, for instance, if the overall death rate in the new data differs from the death rate in the data used to generate the risk model. To the third method's model, the fourth method adds hospital fixed effects, that is, one extra term per hospital except for the one hospital chosen to be left out (a necessary step in this method). Rather than add up the O and the E to get the SMR, this approach gives *odds ratios* rather than relative risks for each hospital directly from the model's output in the same way that it gives out odds ratios for the risk-adjustment variables. The odds ratio for the left-out hospital can be obtained from first taking the negative sum of the log odds of the others and then exponentiating.

7.4 Random Effects–Based SMRs (*)

The fifth method that DeLong et al. (1997) consider involves *random* rather than *fixed* hospital effects. There is a quasi-philosophical distinction between these two types. Fixed effects are typically for when there are a set group of hospitals and we're not interested in any others. This would be the case when comparing, for instance, all six hospitals in a given region with each other. In contrast, with random effects we assume that the set of hospitals in our data represents a *sample* from a much larger population of hospitals. We might want to compare each of those six hospitals in that region with all the other hospitals of a similar type in the country as a whole. This distinction can apply to other factors in the model. For instance, if we include age group in the model, we'll fit age using fixed effects because there's a distinctly limited number of age groups in existence. For hospitals, this may or may not be the case, though one could argue that it is *always* possible to consider hospital effects as random expressions of some underlying process (Ash et al. 2013). With random effects, we assume that we draw the observed hospital effects from a sample with some specified distribution, typically a normal. In other words, the hospital-level death rates are normally distributed. This model can be specified in different ways (Bayesian, likelihood and quasi-likelihood), but the statistical preferences between methods are usually "more of philosophical than practical importance" according to gurus Goldstein and Spiegelhalter (1996). Since their seminal 1996 paper, there have been many comparison studies: more on those in Chapter 9. This random effects model in its simplest form is given by

$$\mathrm{Log}\left(\frac{p}{(1-p)}\right) = \underbrace{\mathrm{alpha} + \mathrm{beta} * \mathrm{score}}_{\text{[FIXED EFFECTS]}} + \underbrace{\mathrm{delta\_j}}_{\text{[RANDOM EFFECTS]}}$$

The "score" is the risk score for each patient based on their age, sex, and so on: young patients with mild disease and without comorbidities might have a score of 0, to which we need to add, say, 1 if they have diabetes, 2 if they're elderly, and so on (see Section 6.8.7 for derivation of a risk score). The delta_j is the random effect for hospital j and is usually taken to be normally distributed with mean zero and SD to be estimated from the data. The set of delta_j's are assumed to be independent of the risk scores. Hospital effect estimates will be biased if these three conditions are met: there exists some patient factor such as functional status that isn't measured in the data, is correlated with the risk score and varies by hospital. In fixed effects models and their resulting O/E ratios per hospital, there will be a bias in the estimate of E. This is when hospitals complain that the risk model doesn't do justice to their patients' sicker-than-average casemix. This bias occurs with random effects models too, but in those there's also bias in the estimate of beta for the effect of the risk score (however, that doesn't matter if we only want to compare hospitals). For more on random effects models, see the readable tutorial by the developers of the MLW in software at http://www.bristol.ac.uk/cmm/software/mlwin/download/manual-print.pdf.

An odds ratio for each unit j can be obtained by exponentiating delta_j. It has the interpretation of the odds of death of patients at unit j compared with the counterfactual odds of death of these patients at an average unit. These are known as empirical Bayes estimates or "shrunken" estimates because the method deals with small sample sizes at some units by "shrinking" their estimate towards the overall mean. Small units will see their estimate shrunk more than large units. In the United States, National Surgical Quality Improvement Program (NSQIP) now uses these shrunken odds ratios, though it was noted that the participants had some difficulty understanding them (Cohen et al. 2013). DeLong et al.'s last three methods are the analogues of methods 1, 4 and 5 but involve deriving the predicted risks from the data at hand rather than an external algorithm.

The odds ratio using delta_j in the aforementioned equation is not the only possible measure of hospital performance from such a model. Mohammed et al. (2012) describe three others in addition to that one, which is the first one they give.

The second one is given by Glance et al. (2003). They state, "Since the random effects component of the hierarchical model accounts for between-hospital variability, only the fixed-effects coefficients were used to calculate the expected mortality rate.... In other words, the fixed-effects portion of the model estimates the probability of death for a patient treated at an 'average quality' hospital." This calculation produces a relative risk rather than an odds ratio because it's the sum of the observed divided by the sum of the expected, with the latter taken from the part of the model other than delta_j and turned into probabilities. It compares the number of deaths at hospital j with the number predicted to occur at an *average hospital*.

The third method is given by Render et al. (2005). They state that the sum alpha + delta_j reflects the log odds of death for the jth provider after risk adjustment. The calculation is slightly more involved than with the previous two, but in essence the resulting SMR compares performance between hospitals for an *average patient*.

Finally, they give the CMS's favoured option (Shahian et al. 2005), the ratio of the "smoothed" mortality to the mortality expected if the provider's patients had been treated by the average hospital. In this approach, the observed number of deaths is replaced by a model-based predicted ("smoothed") number of deaths. The aim of this is to deal with sampling error in the observed number of deaths at hospital j. A small hospital's estimated performance is therefore moved (it's "shrunk" as mentioned earlier) towards the average, such that the SMR at very small ones will almost equal the average. A clear drawback of this is that small units are not likely to be average, as pointed out by Greenland (2000) and a large literature on the relation between the numbers of surgeries performed at a unit and postoperative outcomes for that unit. The article then tries to empirically compare the four SMRs with real (not simulated) data for neonatal units but is greatly impaired by small numbers of deaths at most centres and a desire to find fault with mortality measures in general.

These four SMRs from random effects models differ subtly in their interpretation, and there's no consensus over which to use. We can be guided by philosophical considerations regarding the difference between fixed and random effects and by empirical studies on the impact of our choice on which units are labelled as "poor performers" (Chapter 9).

To further complicate matters, multilevel modelling is not the only valid approach to this problem, as we now see.

7.5 Marginal versus Multilevel Models (*)

When there are two levels to the data, for example, when patients are clustered only within hospitals, a viable alternative to hierarchical modelling is the marginal model. This was briefly mentioned in Chapter 3 but is worth a bit more space here. Marginal models use generalised estimating equations (GEEs) rather than maximum likelihood to estimate their parameters. In addition to the usual assumption that the outcome variable follows the appropriate distribution such as the binomial for binary outcomes, it's assumed that patients within a unit have correlated outcomes but patient outcomes in different units are independent, that is, have zero correlation. One nice feature of GEEs is that we don't need to know the structure of the correlation between patient outcomes at the same unit. The GEE method includes the calculation of robust estimates for the standard errors of the regression coefficients that give the correct results even if the correlation structure you choose isn't the right one or if the strength of the correlation between patient outcomes within the same unit varies somewhat from patient to patient. They tend to be more popular in epidemiology than in health services research.

The interpretation of effect estimates differs for marginal and multilevel models for binary outcomes. This is true not just for patient factors but also for hospital-level effects such as SMRs (Sanagou et al. 2012). The predicted probability of the outcome depends on both patient-level and hospital-level covariates. Let's first think about patient factors and what their coefficients mean. In an ordinary logistic regression model, that is, one that ignores any clustering and is neither marginal nor multilevel, suppose that the odds ratio for being female is 1.2. This means that females have a 20% higher odds of death than males in the population as a whole. It doesn't depend on what hospital they went to. The interpretation of fixed effect coefficients, for example, the effect of being female in this example, is the same in ordinary and marginal logistic regression and is straightforward. In a marginal model, odds ratios estimate the effect of predictors on the population as a whole, averaged over the set of delta_j, rather than their effect on a typical hospital (Hu et al. 1998). In a multilevel model, however, the probability of the outcome for a given patient depends not just on their casemix factors like age and sex but also on the random effect of the hospital that they attended (delta_j). Therefore, the usual odds ratio interpretations, such as females having a 20% higher odds of death than males, apply for comparisons of patients only *within the same hospital* (or those with the same value of the random effect delta_j).

Now let's think about what coefficients for hospital-level factors mean. Odds ratios for hospital-level factors, for example, teaching status, in a marginal model represent the odds of the outcome for hospitals with one value of the factor compared with hospitals with another value of the factor. Again, this is straightforward. In multilevel models, however, this is true only for hospitals with the *same value of the random effect*. This awkwardness is why the 80% interval odds ratio (IOR-80%) was developed. This is the interval around the median of the distribution that covers 80% of the hospital odds ratios. To interpret effects of hospital-level risk factors more generally, the unexplained between-hospital variability needs to be taken into account. See Sanagou et al. (2012) for details of the calculation and a fuller discussion of the differences between marginal and multilevel models when comparing hospital death rates.

Which should we use? As usual, it boils down to which is the most appropriate, given the aims of the analysis. Sanagou et al. (2012) conclude that marginal models are preferred if the patient-level factors are of most interest, whereas multilevel models are preferred for provider profiling. As our aim is clearly the latter, we'll not mention marginal models further.

7.6 Which Is the "Best" Modelling Approach Overall? (*)

This all raises the obvious question of which one is "best" or "should" be used, even if we decide to drop marginal models as just discussed. For process and structural quality measures, none of this applies unless for some reason it's considered desirable to adjust for risk. For outcomes measures, I think there are two main decisions: to use internal or external risk adjustment and to shrink or not to shrink.

Regarding the first, the gold standard is considered to be an externally validated risk adjuster. This means that it was developed in one data set and successfully tested on another. For it to be truly external, the data and the healthcare units used to develop the risk-adjustment algorithm would differ from those in your monitoring scheme. Such a risk adjuster would have been shown to work well in different settings. This gives it kudos, which will help with acceptance among users. It also suggests that it will work well in your setting, though this must be tested. However, internally validated risk adjusters are much more common and, if done well using the principles described in the previous chapter, will be at least as good. They are of course more relevant to your population than any external algorithm. Whether a risk adjuster works in another country or with another set of units is immaterial. If users are involved in the risk-adjustment process, it will help with buy-in. The decision may in any case be settled by data availability.

The second question is less straightforward to answer. With growing interest in Bayesian methods for provider profiling and with the adoption of such models by influential U.S. bodies such as the CMS for their publicly reported measures and by NSQIP, it's worth discussing in more detail (see the Appendix for a layman's guide to the difference between Bayesian and non-Bayesian methods). Cohen et al. (2013) list three main advantages of "shrunken" estimates (sometimes known in the U.S. literature as reliability-adjusted estimates) and multilevel models in general:

1. Multilevel models more appropriately address within- and between-unit variation than single-level models.
2. O/E ratios from single-level logistic regression are derived from subsets of units and so involve more hypothesis tests than multilevel models, whose SMRs are derived from data for all units, thus mitigating multiple testing.
3. Small units will have unrealistic (e.g., 0 or huge) SMR estimates, whereas shrinkage makes what the proponents consider to be more realistic estimates.

Let's consider these points in turn.

First, multilevel models regard the observed data as having been drawn in two stages: (1) patient observations within a unit and then (2) the unit effects from a larger unobserved population of units. At each stage (level of the model), there's random and systematic variation. Single-level models ignore the random variation in unit effects and can yield confidence intervals that are too narrow, thereby falsely labelling some units as outliers. The proportion of hospitals that are rightly and wrongly labelled as outliers under different models will be considered in detail in Chapter 9 as the picture is much less clear-cut than this statement suggests.

The second point is probably not obvious. The empirical evidence for this claim seems to come from simulations by Thomas and Longford (1994) that compared the common O/E ratios from single-level models with empirical Bayes estimates and noted the coverage properties of each set of confidence intervals (CIs). For the 2000 simulated hospitals, the 95% CI for the common O/E ratios did cover the required 95% of rates used to generate the simulated sample and did better than those for the empirical Bayes approach. When the authors looked at the coverage for the 50 hospitals with the highest rates, however, the CIs for the common O/E covered only 24 of these 50 hospitals compared with 48 of the 50 for the empirical Bayes CIs. The authors argued that attention is usually focused on such extreme units, so the coverage properties of the SMR's CI are particularly important. Elsewhere, Greenland and Robins (1991) wrote that the "cost" (by which they mean false positive rate and its consequences) of using empirical Bayes estimates is expected to be less than that of using single-level model O/E ratios; they also give references to the maths and logic behind this. They refer throughout to the empirical Bayes approach as an adjustment for multiple comparisons. Loosely, this has to do with the claim that hierarchical models lead to fewer units falsely classified as outliers (again, see Chapter 9).

For these reasons, NSQIP now report odds ratios from their hierarchical model. They note in one line the controversy around assigning average performance to a small unit, a consequence of the third point. As this is a key feature, it's worth a little more discussion. An alternative approach to assigning average performance is first to estimate the volume–outcome relation for the patient group and outcome of interest and then to shrink the unit's estimated performance not to the overall average but instead to where the volume–outcome relation predicts it would perform (Staiger et al. 2009). This replaces one not very reasonable assumption with one that sounds more plausible, though of course we're guessing. As Dimick et al. (2012) acknowledge, there's an inherent trade-off here. With CMS's shrunken estimates, small units are assumed to have average performance unless proven otherwise – innocent until proven guilty. With these volume-adjusted estimates, they are assumed to have worse-than-average performance until proven otherwise – guilty until proven innocent. There's no "right" answer here. The context is really important. As stated in the report by the Committee of Presidents of Statistical Societies, which was asked by CMS to address criticisms of their hierarchical models, "The committee identifies as a top priority evaluating the option of extending the model to include shrinkage targets that depend on hospital attributes" (Ash et al. 2012). An example of this in a study of how many Ontario hospitals were flagged as high- or low-mortality outliers following an AMI (Austin et al. 2004) was the inclusion of hospital-level characteristics such as size but also factors such as rurality and the presence of facilities for cardiac revascularisation. The inclusion of hospital size in the risk-adjustment model cut the number of outliers from seven to one. Is this an appropriate adjustment? We saw from the previous chapter that we should only adjust for factors that are not within the control of the unit. Hospitals can't change their location and probably can't change their volume of patients or some types of services they provide (at least not easily or cheaply). The authors raised the natural question: should similar outcomes be expected for large, urban, tertiary referral centres and small, rural hospitals, given that they differ substantially in terms of resources and experience? If the answer is no, then we should consider adjusting for these factors in the risk model or choosing more appropriate peer groups than simply the national average. Of these two options, stratification by type of hospital is in general preferred. Another argument against adjusting for hospital factors relates to patient choice. Suppose that you live near two hospitals, one big and one small, and you as a patient want to know which to go to for a particular specialty that both offer. Let's say that this specialty is affected by the volume–outcome relation so that patients who go to larger centres have better outcomes than those who go to smaller ones. Adjusting for hospital size will make the small one look better and the large one look worse, thereby reducing the apparent difference in quality between the two. This won't be useful to you because it obscures the real difference in quality.

TABLE 7.1

Summary of the Main Types of *Standardised Mortality Ratio* Derived from Logistic Regression or Standardisation with the National Average as the Benchmark Unless Specified Otherwise

SMR Type	Numerator	Denominator	Main Specific Assumptions and Issues
Non-hierarchical models			
Indirectly standardised	Observed deaths at unit	Expected deaths after applying national rates to unit's patients	Proportionality: All stratum risk ratios are equal. Simpson's paradox may apply.
Directly standardised	Expected deaths nationally after applying unit's rate to national patients	Observed deaths nationally	Sidesteps proportionality (unless comparing two units' SMRs directly) and Simpson's paradox; numerically unstable if strata are small.
Indirectly standardised using logistic regression	Observed deaths at unit	Deaths predicted from either external or internal risk model	Risk model is suitable for and (re)calibrated to units' data. Proportionality. Simpson's paradox may apply.
Odds ratio directly from logistic regression	Odds of death at unit	National odds of death	Same as above.
Hierarchical models			
Exponentiating the random effect delta_j (DeLong et al. 1997)	Equivalent to odds of death at unit	Equivalent to counterfactual odds of death of the unit's patients at an average unit	These ORs are shrunken towards national average, so small units are assumed to be average.
Using only the fixed effects in the predicted risk (Glance et al. 2003)	Observed deaths at the unit	Expected deaths, derived from fixed effects part of model only	Compares number of deaths at unit with number predicted at average hospital.
Including not only the random effect delta_j but also the intercept alpha (Render et al. 2005)	Formula involving intercept and random effect	Formula involving intercept and random effect	Compares risk of death for an average patient, thus assuming that the unit has average patients.
CMS's SMR (*reliability adjusted*) (Shahian et al. 2005)	Smoothed observed deaths at the unit	Expected deaths	Small units assumed to have average death rate, which is unlikely.
CMS's SMR but adjusting for volume effects	Smoothed observed deaths at the unit according to its volume tier	Expected deaths as above	Small units assumed to have average rate for small units, that is, above-average death rate.

As a summary, Table 7.1 lists the key features of the various types of "SMR" given earlier, which includes odds ratios.

In the previous chapter, we saw how complex a task risk adjustment is, but now we've seen how it's just one element in the larger job of producing risk-adjusted indicators by unit.

7.7 Further Reading on Producing Risk-Adjusted Outcomes by Unit

Risk adjustment is a subject that exercises practitioners and users of the outputs alike. The associated literature is vast and yet the task remains "both an art and a science" with plenty of grey areas. If you are less comfortable with algebra, then look no further than the fourth and apparently final edition of the book edited by Iezzoni (2013), especially if you're after United States–based databases and applications. Section 12 of that edition also covers comparing outcomes across providers using O/E ratios and random effects models. A shorter guide that emphasises risk prediction rather than risk adjustment can be found by reading the *BMJ* series of articles on prognostic models cited in the previous chapter (e.g., Royston et al. 2009).

8
Composite Measures

Aims of This Chapter
- To define composites and outline why they might be useful
- To describe how to construct them, with examples
- To discuss the pros and cons of using composites

What do we mean by composite measures? Here are two typical definitions.

> A composite performance measure is defined as the combination of two or more measures into a single score to simultaneously represent multiple aspects of provider performance and to facilitate comparisons.
>
> **American College of Cardiology Foundation/American Heart Association Task Force on Performance Measures (Peterson et al. 2010)**

> A composite indicator is formed when individual indicators are compiled into a single index on the basis of an underlying model. The composite indicator should ideally measure multidimensional concepts which cannot be captured by a single indicator.
>
> *OECD Handbook* **(Nardo et al. 2005)**

The second adds a bit more to the first by referring to an underlying model and multidimensional concepts. Both of these ideas fit well with quality of care, which therefore seems particularly ripe for the use of composites. Regulators and accreditation bodies need to judge organisations from multiple angles, which means having multiple indicators of quality. Report cards and dashboards are common in many countries, which can soon get crowded with individual measures. Correlations between measures of quality are often weak, and entities as complex as healthcare providers rarely perform equally well in all aspects. As the amount of information on healthcare performance grows, so does the need to summarise it into something digestible. This has long been acknowledged in other sectors such as the hospitality industry. Summaries can be made of only a few items. For instance, TripAdvisor overall ratings for hotels combine ratings for just six aspects such as location and service into a simple average; as we'll see, such averaging is just one way of combining (weighting) information. The public are used to using summaries such as star ratings when choosing things like hotels and restaurants, and there are a growing number of such summaries in healthcare aimed at, and created by, the public (see Chapter 11 for more on public reporting).

In addition to having a potential role in patient choice and in benchmarking providers, composites can also be useful in quality improvement (Profit et al. 2010). Focusing on an individual metric may lead to piecemeal rather than system-based efforts. The latter, such as in the area of patient safety, may have added benefits if they spread within or beyond the institution, such as improving safety attitudes among staff (Sexton et al. 2006).

However, some people believe that composites should be avoided altogether. They question the meaningfulness of a single number or rating resulting from the combination of complex information and the often arbitrary way in which all that information is combined. These are both valid concerns. Can an individual practitioner, clinic or hospital be adequately rated with just one statistic? The debate may well rage forever. As with statistical models and any single indicator, the justification for a composite indicator lies in its fitness for the intended purpose and in peer acceptance (Rosen 1991). If a measure, however flawed technically, is used and leads to improved net performance once any adverse unintended consequences are also considered, then it can be said to have value. If it's not used, if it doesn't help improve performance, or if it causes too much grief for the benefit it delivers, then it's worthless – however good its statistical properties might be.

Now that composites have been defined and the rationale for their use set out, this chapter will first give some high-profile examples of composites, examine the typical steps in their construction and show how our example composites were constructed and validated. We then list the pros and cons of using them before giving some alternatives beyond simply retaining all the individual indicators. We conclude with some references for further, more detailed reading. This is a growing area of research, and some of the statistical methods are detailed and difficult. We therefore present only the main steps and issues – but sufficient to create a solid composite.

8.1 Some Examples

As noted in Chapter 4, the Agency for Healthcare Research and Quality (AHRQ) produces a set of patient safety indicators (PSIs) that are in use in various countries and include such adverse events as infections, pressure sores and obstetric tears. AHRQ combined some of these (AHRQ 2008), and Centers for Medicare & Medicaid Services (CMS) plans to use this composite measure ("PSI 90") in its 2015 Inpatient Prospective Payment System VBP program. The Leapfrog Group of U.S. hospitals combined some of the AHRQ PSIs with survey results and other information to produce an alternative patient safety composite. Both these examples will be outlined later in this chapter.

National regulators and accreditation bodies typically need to combine an even greater amount of information than that just relating to safety. For example, the United Kingdom's national regulator, the CQC, pools a very large number of indicators when prioritising hospitals for inspection (Bardsley et al. 2009).

8.2 Steps in the Construction

Table 8.1 lists the main steps in the construction of a composite and the key issues around each one. These have been distilled from several detailed handbooks or discussion documents (Jacobs 2000; Nardo et al. 2005; Profit et al. 2010).

TABLE 8.1

Main Steps in Composite Indicator Construction

Step	Key Issues
Specify the purpose and choose a framework	For judgment or improvement? Inputs, processes or outputs? Which dimensions of quality?
Choose the units to be assessed	Boundaries are often blurred. Heterogeneity can influence the choice of indicators included and is a greater problem when comparing countries.
Select the data set(s) and indicators	Trade-offs between availability and desirability and between thoroughness and parsimony. Imputation may be necessary if missing data common.
Normalise and scale the indicators	Indicators need to "point the same way", for example, high always means good, while low always means bad. They need to be on the same scale: transformation may be necessary.
Weight and aggregate	Choice of weights (relative importance) will affect units' apparent performance. Need to decide whether units can compensate for poor performance on any indicators.
Conduct sensitivity analyses	The modelling assumptions, data quality, indicator choice, etc., can all affect units' apparent performance. Deconstruct the composite to look for redundancy and performance patterns.
Present results to stakeholders	Method of visualisation vital to ensure that people actually use the results – see Chapter 11.

8.2.1 Specify the Scope and Purpose

Before we can choose our indicators to go into the composite, we must specify the scope of what we are trying to measure, for example, inputs or outputs? As Nardo et al. (2005) stated, what is badly defined is likely to be badly measured. This process should ideally be based on what is desirable to measure and not on which indicators are available, but of course in practice there will be a trade-off here. It may be helpful to define the subdimensions at this stage, so for quality of care this might be the six Institute of Medicine dimensions. It's vital to get expert and user input here where possible, partly in order to maximise buy-in and engagement.

8.2.2 Choose the Unit

Choosing the unit to be assessed arose in Chapter 3. Boundaries between units in healthcare are often blurred as patients are transferred between hospitals or between hospital and social care. If units are too dissimilar, then analysis should be stratified by unit type, for example, by hospital size (see also Chapter 9 regarding choosing the benchmark).

8.2.3 Select the Data and Deal with Missing Values

Our data from which to derive the individual indicators can come from administrative sources, clinical registries, surveys, audits or more ad hoc sources. Chapter 5 described the options so we won't repeat their relative merits here in any detail. Data from any of these sources for a given case (e.g., a patient) can be missing in a random or non-random fashion.

The least bad scenario is when they are "missing completely at random", in which missing values do not depend on the variable of interest (outcome measure) or on any other observed variable in the data set. See Section 6.6 for the options on handling missing data such as dropping the cases and imputing values.

Data may be missing for the whole indicator for a given unit. This isn't necessarily random. For instance, with submissions to a safety incident reporting system, low numbers of submissions is generally considered to represent poor safety culture rather than a genuinely low number of safety incidents. Gaming is also possible here. It's worth doing sensitivity analyses to assess the impact on the affected unit's rating of, for instance, excluding the indicator from the composite for all units. In such analyses, we can assume firstly that the unit has average performance and secondly that the unit has poor performance. Producing results based on the latter should encourage the unit to submit data next time.

8.2.4 Choose the Indicators and Run Descriptive Analyses

We considered issues around the choice of single indicators and their definition in Chapter 4. Let's assume that we have a number of valid, timely, relevant and accessible measures to pick from and have dealt appropriately with any missing values. It's useful to do some descriptive analyses of the distribution of performance across units on each indicator and of the degree of correlation between indicators. An indicator that correlates very highly with another will supply redundant information. At this stage, we may want to trade parsimony against factors such as external validity and thoroughness. On the one hand, we want to capture quality of care as fully as we can with the data available and using indicators considered important and valid by expert evidence and stakeholders. On the other hand, we don't want to include more measures than necessary.

In general, measures with wide a variation across units will yield high statistical power for discriminating among providers. To see how to assess the amount of variation between units in general, see Section 9.4. Specifically with regard to composites, several approaches have been tried. An easy estimate of the variation between units is the interquartile range. This is a natural choice because it's common to rank units and divide into four equal groups or tiers. The drawback is that many outcome measures in health, such as mortality and readmission, are subject to a good deal of apparently random variation. Small sample sizes per unit may also give the impression of variation when in fact little exists. A much more sophisticated approach is to account for the random variation in estimated unit performance by using multivariate random effects models (O'Brien et al. 2007). Here, all the indicators are modelled at the same time (hence the "multivariate" in the name). This model assumes that the units' performance follows a (usually normal) distribution and so will vary randomly around an overall mean (see Chapter 7 for hierarchical modelling and a discussion of random effects). The variation across units in terms of their "true" performance with the random variation dealt with can again be summarised using the interquartile range. A slightly simpler alternative was used by Kaplan et al. (2009). They estimated the intraclass correlation coefficient (ICC) from mixed models for each candidate indicator and only included in the composite those with an ICC of at least 0.3, which seemed to work well.

Sometimes, one measure is of greatest interest or perceived importance, such as 30-day risk-adjusted mortality. It may, however, have some key weakness such as low frequency, and it will of course only represent one aspect of quality. It's interesting to see how our composite compares with such a measure. For example, Chen et al. (2013) did this for three common medical conditions for which mortality rates are publicly reported in the United States for Medicare patients via the Hospital Compare website. They found that their composite explained 68% of variation in future acute myocardial infarction (AMI) mortality rates versus only 39% for the current period's AMI mortality rates. This also demonstrates the large amount of random variation in such outcomes.

8.2.5 Normalise the Metrics

Metrics with different units and scales, for instance, length of stay and standardised mortality ratios, can't simply be aggregated. They need to be transformed (normalised) to a common scale first. There are various ways to do this. The most common ones (Profit et al. 2010) are as follows:

- Rank the units' figures
- Use the distance from some reference, for example, target
- Assign levels on a categorical scale (e.g., star rating)
- Use z-scores

Ranking is the easiest and is not affected by outliers, but it ignores absolute (rather than relative) variations in, and indeed levels of, performance (see Chapter 11 for more on ranking). For national, government or other targets, some measure of each unit's distance from that target can be used. If the target is the top performer, then that top-performing unit receives a score of 1 and the others are given scores of percentage points away from that top unit. This would assume that the unit really is the top performer and doesn't just appear that way due to random variation or data errors.

Categorical scales can be stars or have labels such as "good", "satisfactory" and "poor", or "fully met", "partly met" and "not met". The scores are often based on the percentiles of the distribution of the indicator across units. While easy to understand, such categorisations of continuous variables inevitably throw away information. If categories are chosen carelessly, for example, by dividing units into several tiers with an equal number of units per tier as mentioned earlier, this can give the impression of wide variation when in fact little exists.

The Care Quality Commission in the United Kingdom combines a large number of different kinds of indicators into z-scores in order to prioritise hospitals for inspection (Bardsley et al. 2009). Z-scores are widely used in statistics. This method converts indicators to a common scale with a mean of zero and standard deviation of 1. Indicators with extreme values thus have a great effect on the composite indicator. If this effect is not wanted, then extreme values can be excluded or the indicator downweighted during aggregation.

8.2.6 Assign Weights and Aggregate the Component Indicators

Now that we've chosen and normalised our indicators, it's time to combine them. Choosing the weights to give to each component indicator is one of the most crucial and potentially controversial tasks. Weights must reflect the importance, validity and reliability of individual metrics. As a provider can rate highly on safety but not efficiency, for example, the weights for safety and efficiency need to reflect clinical and policy priorities. Table 8.2 provides the most common options.

TABLE 8.2

Common Options for Weighting and Combining Component Indicators in a Composite

Weighting Option	Description	Comments
Equal	Each indicator gets same weight, so considered of equal importance.	This is the simplest and is a good option if no consensus exists, and many authorities recommend this. This may multiple-count if indicators are highly correlated.
In proportion to data and/or statistical quality		This improves reliability, but gives more emphasis to indicators which are simply easier to measure and obtain rather than important.
Opportunity scoring or denominator-based weights	Some patients are ineligible for some measures. Every triggering of an indicator – "opportunity for care" – has equal value in calculating the composite.	Commonly triggered indicators will swamp less often triggered ones, so they work best if units serve fairly homogeneous populations.
Numerator-based weights	More weight is given to common events than to rare ones.	This may be more sensible for adverse events than for positive events.
Benefit of the doubt	This allows each unit to downweight indicators where they perform worst, so units may all have different weights.	Not used so far in healthcare. Could be used if no indicator is considered indispensable.
All-or-none	The management of a patient is considered a "success" if all the indicators triggered by that patient were met. This avoids choosing weights for each indicator.	It aims at excellence; probably too harsh unless a good outcome is dependent on *all* processes being met.
70% standard	This is a less stringent version of the all-or-none method, where the criterion for success is that 70% or more of the triggered indicators be met.	Seventy percent (or other) threshold is arbitrary. This and the all-or-none method collapse data into pass/fail, which yields lower statistical power.

Weights can be chosen using statistical analysis (considered at the end of this section) or by asking stakeholders. The latter often uses some modified Delphi process (there are other so-called participatory methods) and has the key advantage of stakeholder buy-in and hence face validity. However, this may not transfer well to other types of stakeholders who were not involved in the process. This approach is common, but not universal: the U.S. Society of Thoracic Surgeons Quality Measurement Task Force considered purely subjective weighting by experts but rejected it as "scientifically indefensible" (O'Brien et al. 2007).

Rather than asking experts or users, the simplest weighting scheme is of course just to give all indicators the same weight and hence the same relative importance. This has been found to result in similar performance assessment as other weighting schemes, particularly with a lot of component indicators, unless differences in weights are very large (Bobko et al. 2007). If there are no good alternatives, many authorities recommend this approach.

There are two common ways to assign equal weights. Some U.S. pay-for-performance schemes just take raw averages of the indicators, for example, the sum of the proportions of eligible AMI patients at the hospital who received the recommended therapies divided by the number of therapies with eligible patients at each hospital. This unweighted average is easy to calculate, though the variance is not. It doesn't yield patient-level scores, which limits further analysis, so it yields a measure of average quality across *processes of care*. If we only have information by unit and indicator, for instance, if we are retrieving publicly available figures from the web from sites such as NHS Choices or Hospital Compare, then this may be our only option.

With patient-level rather than unit-level information, instead of averaging across indicators we can average across *patients* so that they are given equal weight in the final score regardless of how many indicators they triggered (i.e., are eligible for). It's been suggested that this is the best choice for evaluating the average quality of care provided to a specific set of patients (Reeves et al. 2007). With both indicator-averaged and patient-averaged scores, the variance depends on both the within-patient and within-hospital correlations, though this is often (wrongly) ignored both here and with the unweighted average described earlier.

If the composite comprises several dimensions and each dimension comprises a different number of component indicators, then the dimensions will in practice generally be given different weights. Particularly, with a small number of components, there's a risk of multiple counting if some components are highly correlated. Developers often respond by testing for this correlation using, for example, Pearson's coefficient, and to choose only those indicators that exhibit a low degree of correlation. Alternatively, they adjust the weights correspondingly, for example, by giving less weight to correlated indicators. Note that there will almost always be some positive correlation between different measures in the same aggregate, so a rule-of-thumb correlation threshold is needed to define when action such as excluding indicators should be taken. There seems to be no agreement on what this threshold should be, though few would argue that a correlation coefficient of at least 0.90 between two indicators does not represent double counting. The main thing is that the threshold is chosen a priori.

Indicators can also be combined giving preference to either their numerator or their denominator. The former gives more weight to common events: in a composite of AHRQ PSIs, pressure sores and infections would drown out the rarer iatrogenic pneumothoraxes and "never events". If the outcome is positive rather than negative, then unit performance is driven by what they do well and what they do well often, at the expense of what they do poorly and/or less often. The CMS recommends giving preference to the denominator, in which they sum all the component numerators, sum all the component denominators and then calculate the ratio of summed numerators to summed denominators.

If patient-level rather than just unit-level data are available, then an "all-or-none" measurement is suggested by Nolan and Berwick (2006) as a way to encourage excellence. This avoids having to give separate weights to each process measure. Providers have to apply all the recommended processes of care to a patient in order to get the credit. For instance, for the 2014 Quality and Outcomes Framework standards for general practitioners (GPs), eight checks were proposed for people with diabetes: body mass index measurement, blood pressure measurement, HbA1c measurement, cholesterol measurement, record of smoking status, foot examination, albumin-to-creatinine ratio and serum creatinine measurement. If the GP does seven of these during the year but not the eighth, it will count as a failure. That might be considered harsh, and critics say that in some scenarios it could lead to perverse incentives such as providing treatment of little value (Hayward 2007).

When aggregating the components rather than using all-or-nothing scoring, another key decision to be made is whether providers should be allowed to compensate for poor performance on one metric with superior performance on another. It's common to allow such trade-offs, but it would be advisable to at least try out the alternative as a sensitivity analysis. An interesting way of implementing such trade-offs is "benefit of the doubt" weighting, in which units can give lower weights to those indicators that make them look bad. It's like students being allowed to downweight or even discount entirely their worst exam result. This hasn't really found its way into healthcare yet.

Weights can also be chosen using purely statistical means, letting the data themselves decide. The most common such approaches are principal component analysis and factor analysis. These methods can be used to group individual indicators according to their degree of correlation. As it does not depend on the views of any particular group of experts, this approach can't be said to pander to those experts' biases. The first step is to check the correlation structure of the data: if the correlation between the indicators is low, then these approaches won't work very well. If there is a reasonable amount of correlation, then these methods create a set of new variables (principal components or factors) that's smaller than the number of indicators. The usual practice is to keep only the most important factors, so those that explain at least 10% of the variance each and at least 60% between them; it's also common but not necessarily best to keep those with associated eigenvalues greater than 1. Factor analysis then usually involves a process called factor rotation (usually the "varimax rotation") to further minimise the number of indicators. Lastly, the weights are derived from the resulting output using one of several available methods.

The main drawback to these statistical approaches is the loss of face validity and stakeholder buy-in. A second one is that weights can't be estimated if the indicators aren't correlated. The efficiency frontier approach from economics is another possible way in which the weights can be determined by the data (see Section 4.5.1). Here, the weights depend on where the unit is located relative to those units on the efficiency frontier. A key drawback of this is that it's not obvious what direction a unit should move in to improve performance.

8.2.7 Run Sensitivity Analyses

Lastly, before the results are presented, some sensitivity analyses should be run. The preceding sections have already pointed out some advisable checks. Developers often distinguish between uncertainty analysis, which looks at how uncertainty in the inputs affects the final results, and sensitivity analysis, which assesses the size of the contribution of the individual source of uncertainty to variations in the final results. Key sources of uncertainty include the selection of individual indicators, data quality, normalisation, weighting and the aggregation method. "Uncertainty" here relates to the best way of doing things or how accurate the data are. For ease, we will just use the term *sensitivity analysis* here. Nardo et al. (2005) give some suggested analyses including:

- Testing inclusion and exclusion of individual indicators
- Modelling data error
- Comparing single and multiple imputation of missing values
- Using alternative data normalisation schemes
- Using different weighting and aggregation schemes

The composite must be able to spot outliers at both ends of the performance spectrum. It's a good idea to deconstruct the composite and examine how its components contribute to the overall score. If a measure contributes little, you may consider dropping it for the reason of parsimony. This decision should be weighed against user acceptability. If the measure contributes little variation but is perceived to be of high clinical importance, then most clinicians may prefer to retain it. An example is mortality in paediatric populations or some elective surgery.

Deconstructing the composite, for example, into domains, can shed light on unit performance: are there units that do well on risk-adjusted outcomes but poorly on access to care? Methods such as path analysis, Bayesian networks and structural equation modelling can help illuminate the relationship between the composite and its components. Structural equation modelling (SEM) is a collection of techniques that allows the relations between the dependent and the independent variable(s) to be investigated: how do the various indicators affect performance on the composite measure? It still often begins with the drawing of a path diagram. Path analysis was invented a century ago and uses a set of boxes and circles connected by arrows to show which variables are measured, which are unmeasured, and the relations between them, whether causal or just correlational. SEM typically continues with factor analysis and/or regression, but we won't consider it further here.

Lastly, it's useful to be able to show a correlation between the composite and other established measures. If none is found, then it's natural to doubt the validity of at least one of the two, and it can act as a spur to check the construction and calculations for the composite.

For technical information on the statistical methods for the aforementioned steps, such as imputation and multivariate analysis, see Section 2 of the *OECD Handbook* (Nardo et al. 2005).

8.2.8 Present the Results

This is covered in detail for performance monitoring in general in Chapter 11, but some issues are specific to composites. While important in general too, transparency in the formulation process is vital to composites for acceptance, particularly if there are influential subjective aspects such as weight choice. Publishing the modeller's objective, methods and results of sensitivity analyses will increase users' trust. Many national and international organisations such as Eurostat, the statistical office of the European Union, have quality standards for their metrics. For example, the European Statistics Code of Practice considers statistical outputs as viewed by users in terms of six quality dimensions: relevance, accuracy, timeliness, accessibility, comparability and coherence (see Eurostat's website for details: http://ec.europa.eu/eurostat/web/quality/european-statistics-code-of-practice).

There is no recognised optimal way to show each of the components together. One option is the stacked bar chart in which each bar represents one dimension and dimension components are stacked on top of one another. The overall leading unit's position could be marked on each bar. Spider diagrams or radar plots are popular, in which the length of the radius of each arc of the spider's web reflects the size of the score on that dimension/indicator (Saary 2008). Colour decomposition diagrams are also popular, especially when there are lots of indicators. "Spie" charts have been advocated, which are combinations of two superimposed pie charts that allow the relative weights to be shown. We'll return to these in Chapter 11 on presenting data.

As mentioned earlier, one weighting scheme is to let users choose. While there are statistical methods to do this so as to optimise a given unit's apparent performance, it would be more flexible and also educational to build an interactive tool on a mobile phone or the Internet that shows users the result of varying the weights. For instance, if a patient choosing a hospital values parking over mortality rates, they could turn the virtual dial to upweight parking at the expense of mortality and see which hospital comes out best. If a healthcare commissioner prioritises efficiency over safety metrics, they could upweight the former at the expense of the latter. Note that we could be motivated to choose weights by considerations of perceived data quality.

8.3 Some Real Examples

8.3.1 AHRQ's Patient Safety Indicator Composite

AHRQ distinguishes between two perspectives on constructing composites: signalling and psychometric. The latter is familiar to us in measuring intelligence – a multidimensional construct – with IQ tests that seek to capture the elements of that construct. People therefore often see parallels between quality of care and intelligence. AHRQ, however, combined their component Patient Safety Indicators (PSIs) from a "signalling" perspective. Their aim is to guide decision making by providing information ("signals") that will result in actions leading to some intended result, that is, safer care. The acid test for their composite is whether such actions are taken, not whether it reflects the underlying construct as would be the case from a psychometric perspective.

Their technical document numbers five steps but really there were these seven:

1. Choose the component indicators (not all PSIs were included).
2. Compute the risk-adjusted rate and confidence interval for each indicator.
3. Scale the risk-adjusted rate using the reference population.
4. Compute the reliability-adjusted ratio.
5. Select the component weights.
6. Construct the composite measure.
7. Evaluate the performance of the composite.

In Step 1, all available PSIs that met the conceptual criteria agreed on by a Working Group were included. This led to a focus on post-procedural adverse events, infections and pressure sores. Step 2 was straightforward. The relative magnitudes of the resulting PSI rates varied across indicators. Therefore, in Step 3, the component indicators were scaled by the reference population rate so that each indicator reflected the degree of deviation from the overall average performance. This consisted of dividing each hospital's PSI rate by the population rate for that PSI to give a ratio. Step 4 addressed the issue of hospitals and also PSIs having very different denominators. AHRQ wanted to give more credence to PSIs based on large numbers than those based on small numbers. For small hospitals, the rate was adjusted to be more like the population mean. This is in line with CMS's "shrunken" risk-adjusted

… *Composite Measures* … 147

outcome rates, in which rates for small hospitals are "shrunken" (i.e., moved) towards the overall mean (see Section 7.4). For Step 5, they considered and presented results for several different types of weight used when combining the components as a weighted sum in Step 6. The calculation of confidence intervals is often complex due to correlations within and between patients, and the technical appendix gives the details. The composite measures were evaluated using three criteria: discrimination between units (how many units were declared to be outliers), forecasting (the ability of the composite to predict the performance for each of the component indicators) and construct validity (the degree of association between the composite and other aggregate measures of quality, which could be other composites). The PSI composite was ultimately endorsed by the National Quality Forum in 2009.

8.3.2 Leapfrog Group Patient Safety Composite

The Leapfrog Group is a national coalition of U.S. healthcare purchasers with a history of well-respected quality improvement initiatives. They wanted to create a single, publicly available safety score for every U.S. hospital that would balance structure, process and outcome. In contrast to AHRQ, who simply had to decide among their own set of indicators (though the selection and definition of each PSI had been an extensive and rigorous procedure), Leapfrog could cast a wider net. They involved nine national experts who weighed up the evidence base for, and relevance to patient safety of, each candidate indicator. The weights were calculated using the following formula:

$$\text{Weight} = \text{Evidence} + \text{Opportunity} * \text{Impact}$$

The experts came up with the values for the elements in the formula for each indicator. Their next task was to aggregate the indicators, which first needed to be standardised to a common scale. They chose the z-score method for this, in which the hospital's performance is expressed as a number of standard deviations from the national mean. This common approach is easy to understand and explain, and it's been used in many other medical spheres. As mentioned earlier, one potential disadvantage of the z-score, or indeed any rescaling approach, is the exaggeration of clinically meaningless differences when measures have a small standard deviation. The Leapfrog group tackled this by giving more weight to measures with higher coefficients of variation (ratios of the standard deviation to the mean).

The expert panelists considered different ways of dealing with missing data for whole measures and had sensitivity analyses run. They chose to exclude the missing measure and recalibrate the weights for the affected hospital(s) using only those measures for which data were available, though they also used imputation with the aid of alternative data sets for two survey measures.

Finally, the composite score itself was calculated by multiplying the weight for each measure by the hospital's z-score on that measure. They added 3 to avoid having negative scores.

Their paper reports variations by hospital characteristics and state and the distribution of scores by degree of missing data (Austin et al. 2014). They considered it a strength of the measure that mean scores only varied slightly by factors such as the proportion of patients insured by Medicaid or by rural versus urban location. They recommended comparison of the scores with outcomes such as mortality and readmission rates and with other safety composite scores.

8.4 Pros and Cons of Composites

Table 8.3 summarises the main pros and cons of composites and is an amalgam of the University of York technical report (Jacobs 2000) and the *OECD Handbook* (Nardo et al. 2005). In short, as you might imagine, composites are

TABLE 8.3

Main Pros and Cons of Healthcare Composite Measures

Pros	Cons
Can summarise complex, multidimensional realities into a rounded view of performance – the big picture – for any kind of unit, including a whole system	May send misleading policy messages or be misused, for example, to support a desired policy, if poorly constructed, presented or interpreted; to be comprehensive, may have to rely on poor data in some dimensions
Are easier to interpret than a battery of many separate indicators and can avoid "information overload" that can lead to poor decisions	The selection of indicators and weights could be the subject of protracted arguments
Facilitate communication with the general public and promote accountability	May invite simplistic policy conclusions
Enable users to compare complex dimensions effectively	May increase the difficulty of identifying proper remedial action
Reduce the visible size of a set of indicators without dropping the underlying information base	May disguise serious failings in some dimensions
Help to focus policy on issues of unit performance and progress	May lead to inappropriate policies and distort behaviour if dimensions of performance that are difficult to measure are ignored
Can more easily assess progress of units over time	Deterioration in one dimension may be obscured by improvement in another
Can give greater statistical power to detect differences between units	May complicate confidence interval calculation

a trade-off between ease of presentation (digestibility) and capturing all the main variation in performance (detail). We'll return to this in Chapter 11. The other main point here is how actionable the results are. As measures become more and more aggregated, it becomes more difficult to determine the source of any poor performance suggested by the composite and hence where to focus remedial action. This is reminiscent of the process–outcome divide, in which a patient outcome may be considered at least partly the result of numerous clinical processes: finding that a unit scores poorly on the outcome does not tell us which processes are at fault.

8.5 Alternatives to the Use of Composites

As we said at the outset, some argue against using composites under any circumstances. You may decide not to combine. For outcome measures such as death and complications, there are two notable alternative approaches.

First, composite endpoints are common in randomised controlled trials. In health services research, the combination of death and readmission into one is particularly common – the single outcome analysed is therefore readmission *or* death (Dhalla et al. 2014; Lau et al. 2015). This avoids the problem of censoring (patients being lost to follow-up in the data) and a potentially negative correlation between the two – patients who die can't be readmitted, and if poor care leads to more deaths it may also result in fewer readmissions. However, it also makes the huge assumption that both outcomes are equally important, which is definitely not true from the patient's perspective. It may also lose information if the predictors and patterns of each outcome differ.

Another option is akin to the "all-or-nothing" combination of process measures described earlier. Outcomes may be combined into a "trouble-free pathway" measure in which success is defined as the avoidance of any of a set of adverse outcomes (e.g., delays, complications, readmission, unplanned reoperation, death; Perla et al. 2013). We find this appealing and expect to see more applications in the literature in future.

Finally, a more statistical approach is to try a latent variable model (Landrum et al. 2000; Teixeira-Pinto and Normand 2008), identical to item response theory (IRT) models used extensively in educational and psychological testing. IRT has also been used in the construction of patient-reported outcome measures (PROMs). These models assume that the set of indicators reflect one single underlying construct (quality of care or, for the PROMs, quality of life) but indicators may have a different weight in the final score, depending on their ability to discriminate between units. Put another way, indicators that vary most across units get the biggest weights. This avoids any subjectivity in choosing the weights, though a priori beliefs about the relative importance of the set of indicators can still be included by using prior distributions in a Bayesian framework. It also deals with the within- and between-patient correlations. However, it assumes that the latent variable is normally distributed, which can't be verified, and there are other theoretical disadvantages. See Teixeira-Pinto and Normand (2008) and Landrum et al. (2000) for more details.

9

Setting Performance Thresholds and Defining Outliers

Aims of This Chapter

- To set out the different ways to compare units in terms of their performance
- To define outliers: units with "unusual" performance
- To outline statistical process control methods
- To discuss how the choice of risk-adjusted outcome measure (SMR) affects which units are flagged as outliers

After deciding what to measure and how to measure it, with adjustments for casemix as necessary, we come to the business end of performance monitoring: deciding which units if any have performance levels that are worthy of further attention. What that attention is depends on the purpose of the monitoring, as has already been stressed. We generally want to know whether a unit's performance is "unacceptable", although P4P schemes, for instance, are also interested in "good" or "excellent" results. A quality improvement project may simply want to know whether care has improved at that unit – even if other units have improved more during the same time period. The drive for "high reliability systems", using approaches borrowed from industry, includes a focus on reducing variation both within and between units. It's therefore useful to be able to quantify the amount of inter-unit variation and monitor it over time. In this chapter, we'll see how to do all of these. A couple of sections are necessarily more technical than the rest and are marked with an asterisk as usual.

There are three principal ways of defining acceptable performance:

1. A priori external targets
2. Historical performance at the same unit
3. Referring to inter-unit variation

We'll now consider each in turn.

9.1 Defining Acceptable Performance

Units not reaching the acceptable level are often called "outliers" – and so are those whose performance is unusually good. There is generally less focus on these "good outliers" in monitoring. Targets, when they are set, more often than not refer to minimum acceptable performance rather than minimum excellent performance. "Good outliers" are therefore usually defined using the third method, for example, by being in the top 10% of units or lying beyond three standard deviations from the average. Although much of our discussion will also relate to spotting excellent units, in this chapter we'll focus on the definition of poor performers as there's good evidence that those are the ones that people and organisations take notice most.

9.1.1 Targets

Targets may be set by international organisations such as the World Health Organisation (WHO), by the government, by the national regulator or accreditation body or by a professional society such as the American Heart Association or the Royal College of Surgeons. The target may be absolute, such as a 0% infection rate, or relate to best practice, that is, what the best units regularly achieve. A unit may then simply be compared with this fixed value using 95% confidence intervals; more conservative ones such as 99.8% may be used to control the false positive rate if there are large numbers of units to be assessed – we will refer to this issue of multiple testing again later in this chapter. If the target is 0 or 100, for example, an infection rate of 0% or process achievement of 100%, then no statistical testing is needed – a single infection will mean that the unit has not met the target. Whether it's fair to demand such levels of perfection is not a statistical decision. For many years, it was considered impossible to achieve a zero ventilator-associated pneumonia (VAP) rate, but a number of hospitals persistently now do so (there are, however, controversies around the diagnosis and definition of VAP). The relevant professional society has a key role here. The Institute for Healthcare Improvement's (IHI's) Pursuing Perfection demonstration program that ran during the 2000s was to "build the Toyota of healthcare", with ambitious goals to exceed previously believed "limits". Such outstanding performance was referred to as the Toyota Specification. The use of a priori targets is an example of what's sometimes called criterion-referenced assessment or comparison against set standards. With this method, it's possible for all units to "pass" or even for all to "fail". This is common in education, where the teacher/examiner wants to know if the

Setting Performance Thresholds and Defining Outliers 153

student has learned enough of the course material. In healthcare, having targets can sometimes give unexpected results. The United Kingdom has the world's largest P4P scheme for primary care, the Quality and Outcomes Framework (QOF), which was introduced in 2004. General practitioners (GPs) are financially rewarded for monitoring and treating their patients with chronic diseases via a series of evidence-based indicators, each with a performance threshold that, when reached, triggers extra payment. The government was surprised – and financially discomfited – in the first year of the scheme when GPs were found to be better at reaching the thresholds than expected. This led to large payouts and a raising of all lower achievement thresholds for existing indicators to 40% in 2006. Targets can have unintended consequences too – see Section 4.11 (in Chapter 4) for how they can distort staff behaviour.

9.1.2 Historical Benchmarks

Quality improvement projects typically compare a unit's new level of performance solely against its historical level (aka ipsative assessment). This is common in plan–do–study–act cycles with statistical process control (SPC) methods (see Section 9.3). In the planning part of the cycle, an aim statement defines the goals for improving performance by a certain amount over a defined time period. Such a goal might be to reduce the infection rate by half within 6 months.

Sometimes, the goal for improving on previous performance incorporates how other units fare. A hospital may aim to improve its standardised mortality ratio (SMR) from 110 one year to 100 the next. As explained in Chapter 7, the SMR is often calculated as the ratio of the observed number of deaths to the expected number within a given time period, with the expected number incorporating adjustment for casemix. A reduction from 110 to 100 therefore represents a fall from 10% higher than the *first* year's average to something equal to the *second* year's average. Suppose that mortality falls over time nationally. The hospital may see a fall in the second year in its number of deaths or in its mortality rate compared with how it did in the first year. However, if the hospital doesn't reduce its mortality rate by *more* than the reduction achieved by the rest of the country, then it won't reach its goal of reducing its SMR because the goalposts have moved from the first year to the second.

9.1.3 Referring to Inter-Unit Variation

Perhaps the most common situation, however, is the lack of accepted target values. We then have to inspect the performance distribution across all units and define an "acceptable" amount of variation, that is, an *acceptable range* of performance. Units falling outside this range are labelled outliers. This is known as norm-referenced assessment, particularly in education where student test results are compared with those of their peers. As we shall see later, there may be different degrees of "unacceptable": values that lie more than 2 SD above the benchmark can be flagged as amber on a traffic light classification system and rates above 3 SD flagged as red (also see Chapter 11 on presenting information).

It's fair to say that although government-imposed targets such as those concerning emergency department or surgery waiting times often generate news headlines, there is more research and debate around methods that rely on looking at performance distributions for deciding what's unacceptable. The choice of peer group has already come up in previous chapters. There's a hoary old joke about two epidemiologists (they're the ones broken down by age and sex), Bill and Fred. Bill says to Fred, "How's your wife, Fred?" To which Fred replies, "I don't know, Bill. Compared with what?" For the purpose of discussion, let's assume that the peer group or benchmark that's being used is the national average. We therefore want to spot units whose performance is "significantly different" from the national average, especially those worse than it.

The simplest approach is to rank the units from "best" to "worst" using just the point estimate of the measure of interest, for example, the unit's death rate for the time period. This used to be done a lot in healthcare and is still common in education with league tables of school exam results. If no confidence intervals are used, then some number of units at the wrong end of the table will be labelled outliers. As with more sophisticated methods, how many units are thus labelled could depend on what resources are available to inspect or penalise (or reward) them. Some P4P schemes divide the units using quartiles, penalise the "worst" quarter and reward the "best" quarter. This is a simple and transparent method, but it has some huge drawbacks. No attempt is made to assess how large the performance difference is between the rewarded and the penalised, which can be pretty small. No account is taken of random variation that results in a unit being the wrong (or right) side of a threshold just due to sampling error, which can be pretty large. As Poloniecki (1998) put it:

> Even if all surgeons are equally good, about half will have below average results, one will have the worst results, and the worst results will be a long way below average.

Beloved of journalists, league tables have long been discredited in healthcare and should therefore be confined to sports.

Table 9.1 summarises these common ways to compare units.

TABLE 9.1

Common Ways to Compare Units

Method	Description	Pros and Cons
Target (criterion-referenced assessment)	Units are compared against an a priori fixed target.	Simplifies analysis, especially if target is 0% or 100%; likely to have better buy-in if target is set by relevant professional body; possible for all units to "pass" or "fail"; targets can distort behaviour and lead to gaming and unintended adverse consequences.
Historical comparison (ipsative assessment)	In quality improvement projects, we usually only want to know whether our unit is doing better than it did in the previous assessment time period.	Comparing against the unit's own previous performance reduces the effect of imperfections of data coding and risk adjustment: if casemix is stable, then crude outcomes can be tracked. Tells us nothing about how we're doing compared with other units or whether we've improved more than they have in the same time.
Peer comparison (norm-referenced assessment)	Perhaps the most common: the national average is the typical comparator. Also possible is the performance by the set of "top performers", e.g., best 10%.	Often, no a priori target can be set. The choice of peer is important: national average may not be fair for tertiary referral or specialist units.
Historical-peer hybrid	In a hybrid form, such as when tracking SMRs which are themselves peer-referenced measures, units are compared with how their performance relative to their peers (e.g., national average) changes over time.	A unit can improve its mortality *rate* without lowering its SMR (see text) if its peers cut their rates by even more.

9.2 Bayesian Methods for Comparing Providers

Rankings in league tables are inherently unstable. This is why lots of studies that use them to compare different indicators or measurement methods inevitably find differences, though many of these will be unimportant in clinical or policy terms. Marshall and Spiegelhalter (1998) improved on this by using a Bayesian approach to take account of the uncertainty in each unit's ranking. Using simulation, their model yields an estimated ranking with its associated credible interval (Bayesian version of a confidence interval) and the probability that a unit is ranked in, say, the worst 10% of units. A unit can then be declared an outlier if this probability is above some arbitrary high threshold such as 80%.

It's quite possible that no unit will exceed this threshold. If this happens, then the model has concluded that there isn't enough real variation in performance to be confident that there are any outliers. One of the great strengths of the Bayesian framework is the ability to make such probabilistic statements as, "there's a 90% chance that this unit is among the worst 10% of units and a 65% chance that it's the very worst one in the country". Classical (non-Bayesian) confidence intervals can't do this, though we frequently pretend that they can when interpreting them. We (AB and PA) like what Bayesian ranking offers and once managed to get them used in the annual publication of hospital death rates (Dr. Foster's Good Hospital Guide) in the national press instead of the old naïve league tables that had been published in previous years' Hospital Guides. What we found, though, was that even with probabilities for each unit's rank position given, people almost inevitably ignored them and went straight to the plain rank.

There is a growing literature on Bayesian methods for provider profiling. As no consensus has yet emerged on which to use, Austin (2002) compared four approaches based on the random intercept model. The first method, proposed by Normand et al. (1997), examines the probability that the hospital-specific mortality rate for a specific patient profile (i.e., predicted risk of death, which Normand et al. took to be the average for the whole set of patients) exceeds a specified threshold. The second method ranks hospitals according to the probability of mortality for a specific patient profile and thus applies Marshall and Spiegelhalter's ranking method to Normand's patient profile method. Both these methods consider only average patients at each hospital. The third method, based on ideas presented by Christiansen and Morris (1997), involves determining the SMR for each hospital and calculating the probability that it exceeds specified thresholds, for example, 15% higher than the average, so if SMR > 115. The fourth method involves ranking hospitals according to their SMR and thus combines elements of Christiansen and Morris and Marshall and Spiegelhalter. Austin found poor agreement between the four approaches in most scenarios tried and called for urgent research into which methods are best able to discriminate between units.

There are two main "versions" of Bayesian analysis: full (or standard) Bayes and empirical Bayes. Both involve combining a priori beliefs about the parameter(s) of interest with the available data when fitting the model; the non-Bayesian way is just to use the data to do this. In standard Bayesian models, the a priori beliefs are decided quite separately from any data. The beliefs are expressed as "prior distributions", each with their own parameters (called hyperparameters) such as their mean and SD that define them. In the empirical Bayes, however, the prior distributions are actually determined from the data. As described in Chapter 7, the Centers for Medicare & Medicaid Services use an empirical Bayes approach to calculate their SMRs and related measures. In multilevel models, empirical Bayes can be seen as an approximation to the standard Bayes. These SMRs take account of random variation within a unit (i.e., sampling error) so that its estimated SMR is adjusted ("shrunken")

Setting Performance Thresholds and Defining Outliers 157

to make it closer to the national average. 95% CIs are then applied, albeit in a necessarily complicated manner due to the underlying assumptions. Outliers are defined simply as those SMRs with CIs that don't include 1.

Jones and Spiegelhalter (2011) discuss how hierarchical models are used to define "unusual" provider performance. They make the useful distinction between units that are "outlying" and those that are merely "extreme". Outliers have SMRs that, following hypothesis testing, are found to be too big to come from the assumed (null) distribution in which all the units perform the same. Extreme SMRs, however, do come from the same distribution as SMRs from all the other units, but this is because the distribution has been made to accommodate them. They are merely far to one end of it. A provider might be considered extreme if its rate has a large posterior probability of being greater than some external target or specified quantile of the distribution. In some studies, units have been declared extreme by comparing against the mean. Jones and Spiegelhalter disagree with this approach. The SMR for a large unit can have a narrow credible interval around its estimated rank and have a high chance of being higher than the mean – and yet not be extreme. For instance, a unit could have an SMR of 105, so just 5% above the national average, with a 95% credible interval of 102–108 that means it's significantly high but hardly extreme. This is the old distinction between clinical and statistical significance, where the latter is influenced more by sample size than importance.

9.3 Statistical Process Control and Funnel Plots

A common definition of outlying performance when units are compared with other units rather than fixed target values is more than three standard deviations above or below the mean. The use of 3 SD limits is a feature of Shewhart charts, devised by Walter Shewhart who revolutionised quality control at the Western Electric Company in the 1920s. Shewhart charts and funnel plots belong to a family of methods in the specialty of statistical process control (SPC). SPC has been described as a philosophy, a strategy and a set of methods for improvement of systems, processes and outcomes (Carey 2003). Its main tools are graphical methods such as Shewhart charts (or "control charts"), run charts and simpler plots such as histograms. We'll see examples of these shortly. Although SPC originated in industry, with its need for quality control of manufacturing and other processes and for the reporting of business performance, it has been vigorously taken up (eventually) by healthcare since around 1990. The IHI and The Joint Commission in the United States have been strong advocates of this approach and have contributed to its spread.

Charts are available for both cross-sectional and longitudinal monitoring. The simplest is the run chart. In this chart, data are plotted at each time point for a given unit's measure, which may be at patient or aggregate level, for

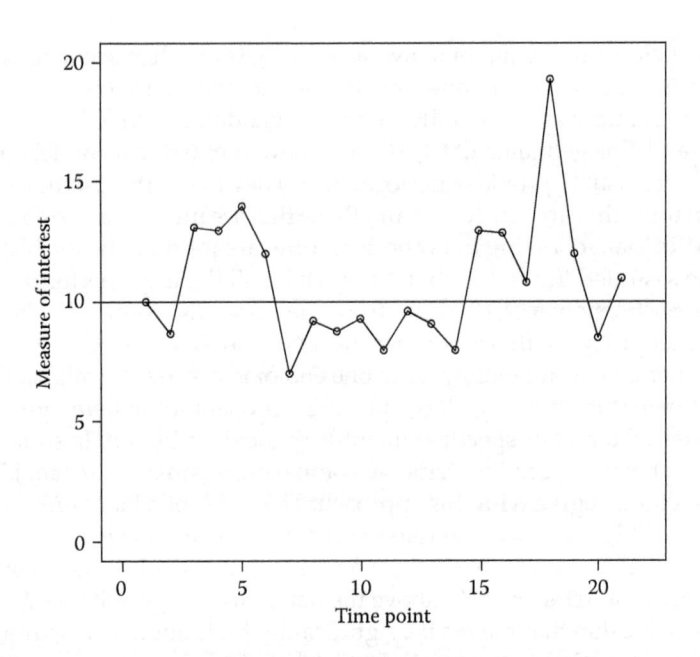

FIGURE 9.1
An example of a run chart showing a run of eight data points below the median.

example, waiting time for each patient or average waiting time for patients seen that week. The median is superimposed, but that's the only statistic that needs to be calculated. A "run" is defined as one or more consecutive data points on the same side of the median. Figure 9.1 shows an example of a run chart, which has the median of the points as its centreline. It has a run of eight points between time 7 and time 14, which would be considered unusual: we'll consider definitions of "unusual" shortly.

Run charts are simple to construct, can be used with any type of data and are useful for monitoring one or maybe a small number of measures within a healthcare organisation. More powerful are control charts, which also plot performance on some measure over time but have upper and lower control limits in addition to the median line. These two extra lines are commonly placed at 3 SD either side of the mean; the mean is used rather than the median as in run charts. Figure 9.2 provides an example of the archetypal Shewhart chart. One data point lies above the upper control limit.

With this type of chart, we need to know the type of data that we're plotting in order to calculate the control limits. Charts exist for means, proportions, counts and standard deviations. If the event of interest is rare, as with healthcare infections, it can be advantageous to monitor the time between events (number of patients who didn't have the event) using the g chart. Some charts are often used in pairs, for example, the x-bar and s chart, which monitors both the mean and the SD of the measure of interest. For the control limits, we also need to use different formulae from those used to calculate the SD of a distribution.

Setting Performance Thresholds and Defining Outliers 159

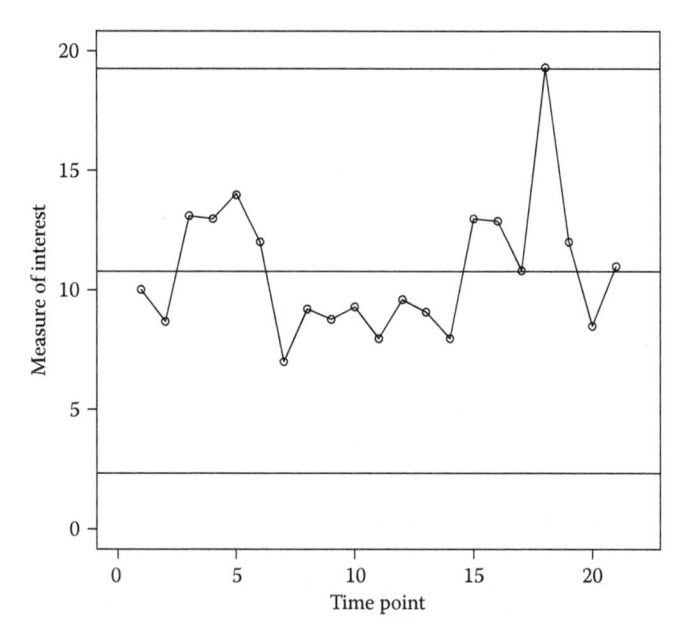

FIGURE 9.2
An example of a Shewhart chart of the data from the run chart of Figure 9.1, with upper and lower control limits with 3 SD on either side of the mean.

Many rules have been proposed to decide when a given chart shows unusual performance. In the language of SPC, this unusual performance is known as "special cause" variation, as opposed to "common cause" variation, which is simply random. Shewhart himself used only one rule: whether a single data point fell outside the control limits, that is, above the upper or below the lower. Figure 9.2 has one such point, at time 18. The Western Electric Company, whose first edition of its *Statistical Quality Control Handbook* was published in 1956, included other rules that have become the standard. An example rule is that special cause variation occurs when eight or more successive points fall on the same side of the centreline (similar to a long run in a run chart) with at least 20 data points in total – our data in the figures include such a run. Another is a trend, defined as at least six or seven consecutive increases or decreases. These rules were designed to minimise false alarms that would lead to the shutting down of an assembly process. In healthcare, however, they could delay the detection of a serious problem that puts lives at risk, and so other rules might be preferred such as using 2 SD rather than 3 SD. When a chart shows special cause variation – for instance, with a long run or when a data point falls outside the control limits – the chart is said to signal. We'll also use the term alarm for a signal, especially in the term false alarm. We'll return to the issue of choosing chart thresholds when discussing cumulative sums (CUSUMs) and related methods later.

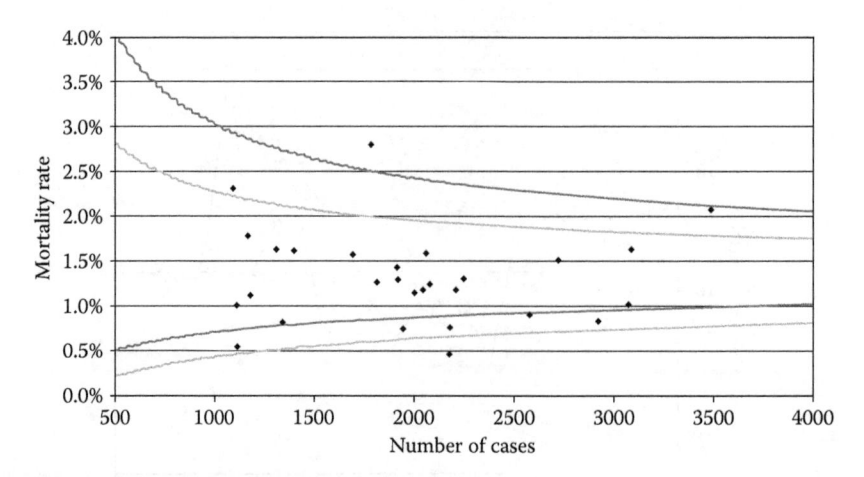

FIGURE 9.3
An example of a funnel plot: inner lines represent 95% and outer lines 99.8% control limits.

Control limits based on 3SD either side of the mean are also standard in the increasingly popular funnel plot (see Figure 9.3). Funnel plots are usually applied to cross-sectional data and seem to be particularly common in surgery (Mayer et al. 2009). They usually have two sets of lines (95% and 99.8% control limits) that together, when seen from side-on, take the shape of a funnel. The inner pair of lines correspond to roughly 2 SD and the outer roughly correspond to 3 SD either side of the mean. The eye is naturally drawn to those units lying outside the 99.8% control limits; each unit's position relative to these limits can be summarised in bands with labels such as "above average", "as expected" and "below average"; labelling and other issues will be covered in Chapter 11.

More complicated than the funnel plot but better suited to longitudinal monitoring are cumulative SPC methods such as the CUSUM, VLAD (variable life-adjusted display, an acronym designed specifically to amuse its creators, so they told me) and EWMA (exponentially weighted moving average). In these charts, the statistic that is plotted at each time point reflects not only the patient(s) on whom the measurement was taken for that time point but also a contribution from all previous patients since the start of monitoring. In the CUSUM, each previous patient contributes as much as the one(s) at the current time point. In EWMA, the contribution of previous patients is downweighted the further back in time you go. With either method, when the chart statistic exceeds a preset threshold, the unit is said to be an outlier. It's common for these charts to plot one patient at a time, and they can therefore be used for up to the minute monitoring, given available data. Risk-adjusted outcomes can be plotted too (Steiner et al. 2000). The chart statistic in VLAD plots is just the difference between the observed and expected outcome for each patient, so for mortality this would be 0 or 1 for the observed and some predicted probability of death for the expected. These are used in Victoria, Australia, for acute myocardial infarction (AMI) mortality (Andrianopoulos et al. 2012).

It's tempting to interpret the cumulative O/E on these plots as "excess deaths" or "lives saved", depending on whether it's positive or negative. Tempting, but hazardous, due to the various errors in sampling and measurement that explain an unknown proportion of the O/E. This brings to mind the noted difficulties in estimating how many lives were indeed saved as a result of IHI's 100,000 Lives Campaign that ran for 18 months from 2004 (Wachter and Pronovost 2006). The chart statistic in log-likelihood CUSUMs, on the other hand, is as a function of the difference between the observed and the expected and is therefore not interpretable as "lives saved" or anything else that makes sense to users. However, these CUSUMs give a lower false alarm rate than other similar charts (Steiner et al. 2000), which makes it valuable for a national regulator. One drawback is that, as already mentioned, all patients in both VLAD and CUSUM plots are weighted equally, so there is the question of how relevant previous patients (and hence previous performance at the unit) are to current patients and performance. This is the same problem with combining multiple years of data into a single cross-sectional measure – units frequently argue that their performance has improved recently and that they can't be judged on a measure that includes data from several years ago. The EWMA addresses this by downweighting patients further back in time, though at the cost of having to set or estimate an extra parameter, lambda.

SPC methods that look for a *change* in performance over time assume that a reasonable amount of historical data is available to establish the in-control (i.e., constant and stable) performance levels. If this baseline performance does change in a regular way, for instance, linearly or seasonally, then time-series methods can be applied to detrend the data so the baseline becomes stable. Time-series data will be considered further in Chapter 12 on evaluation. What can we do if we don't have any/enough baseline data? A Bayesian approach has been suggested that tests for the existence of a change point rather than deviation from some historical baseline. Model selection is done using the Bayes factor. The method appears to warrant further investigation, though the simulations were restricted to a single surgeon's data (Zeng and Zhou 2011).

As regression has its residuals, R^2 and c statistic among others, so SPC has its own measures for describing its statistical performance. These are mostly based on the time each chart takes to signal and whether the signal is real or just a false alarm. The technical literature tends to focus on the average run length, which is the average number of time (data) points that the chart takes to signal since its beginning. We can compare this when the process being monitored is "in control" (i.e., there is *no* special cause variation, so any alarms will be false alarms and we *don't* want the chart to signal) with when it is "out of control" (i.e., there *is* special cause variation, so we *do* want the chart to signal). We clearly want the former to be long and the latter to be short. As the run length distribution is commonly skewed, an average isn't an ideal summary measure, and there are other limitations (Frisen 1992). Also, most people are more familiar with the concept of the false alarm rate, which here is the probability than an in-control chart will signal within a given time. Note that with SPC, all charts will eventually signal, so chart performance measures need to have their timeframe specified. The converse of the false alarm rate is the successful detection rate, which is the probability that an out-of-control chart will signal within a given time. These rates are typically and most easily estimated by simulating a large number of in-control or out-of-control artificial units. Where we set the control limits or the chart threshold is a trade-off between keeping a lid on the false alarm rate while maintaining a reasonable successful detection rate. Where the sweet spot is or what is considered "optimal" depends very much on the context, the cost of both false alarms and failing to detect real changes and the type of method used. The two chart performance measures just described are also not the only ones proposed or in use, but they are among the most easily interpretable. Technically minded readers should see, for instance, Frisen (1992) for a review of performance measures and Frisen (2003) for a way to compare different definitions of optimal.

Table 9.2 summarises the various ways of defining outliers that we've discussed in this chapter.

TABLE 9.2

Common Definitions of Outlying Performance

Definition	Method Used	Comments
Worst-ranked X performers	League tables	Simple and still common, e.g., in education; takes no account of random variation or the size of difference in performance between units; also see Chapter 11; consign to history/sport.
Units with at least probability P of being in the worst-ranked X performers	Bayesian ranking	Explicitly acknowledges that ranks are unstable; need to specify P and X; can disappoint end users who expect/want to see lots of outliers, of which there are generally few with common sample sizes.
Outside preset limits for "acceptable" performance	Simple descriptive statistics	Sufficient if "acceptable" refers to, e.g., zero infections. Key is of course to define acceptable in a way that's acceptable to stakeholders.
95% CI not including benchmark value	Confidence intervals	Familiar to most users; susceptible to false positives if multiple testing (Section 9.4).
3 SD above/below the average	Funnel plots, Shewhart charts or similar	Simpler to calculate than Bayesian ranking. 3 SD threshold is arbitrary but has some mathematical backing. As with Bayesian ranking, users may be disappointed if few outliers are seen. There may, however, be lots of outliers (see text) – what to do then?
Above a preset threshold	Cumulative SPC methods such as CUSUM, VLAD and EWMA[a]	Threshold typically derived from simulation and explicitly deals with the problem of testing at multiple time points; more complicated to implement.

[a] SPC, statistical process control; CUSUM, cumulative sum; VLAD, variable life-adjusted display; EWMA, exponentially weighted moving average.

9.4 Multiple Testing (*)

A final point to make in this brief summary of cumulative SPC methods refers to setting the threshold. This is, as so often, a trade-off between sensitivity and specificity. A lower threshold gives earlier detection of potential problems at the cost of more false alarms: raise the threshold to reduce the false alarm rate but slow the detection. One common way to set the threshold is to use simulation to achieve a desirable false alarm rate such as 1 in 1000 in a year of monitoring. This can be further tailored to the size of the unit and the average outcome rate (Bottle and Aylin 2011b), both of which affect the chances of false and true alarms. Bigger units treat more patients and hence have bigger sample sizes, yielding greater statistical power to detect deviations in performance than with smaller units. Regarding the average outcome rate, a doubling in the rate of a rare event is harder to spot than the doubling in the rate of a common one.

This mention of alarm rates brings us to the crucial topic of multiple testing, which appears in many subareas of medical statistics. It's the main reason why randomised controlled trials have primary and secondary endpoints, when the success or failure of the treatment of main interest is judged on one or a very small number of primary endpoints rather than the usually much larger number of secondary ones. Each statistical test associated with each endpoint has a false alarm rate. This rate must be controlled to avoid concluding that the treatment of main interest is superior to the alternative(s) when it isn't. The best-known way to control the overall ("familywise") false alarm rate is the Bonferroni correction. If a trial is testing four hypotheses that each have a desired false alarm rate (also known as the type I error rate, or alpha) of 0.05, then the Bonferroni correction would test each individual hypothesis at alpha = 0.05/4 = 0.0125. To be deemed statistically significant, the p value now needs to be less than 0.0125, not just less than 0.05. The correction is notably conservative and can miss genuine differences between groups. It's most useful when there are a fairly small number of comparisons to be made and you're looking for one or two that might be significant. With a large number of comparisons and many that might be significant, however, the Bonferroni correction can give you a lot of false negatives. It controls the probability of at least one type I error in the set ("family") of hypotheses being tested, hence the name "familywise error rate". Another problem with it is that there's no consensus on how to define a family. The advice is to consider just how bad getting a false positive would be for your study and so to choose the families a priori (McDonald 2014).

A more recent and promising option is to consider the false discovery rate (FDR), which is less conservative than the Bonferroni correction and its modified alternatives. This is the "expected proportion of errors among all rejected null hypotheses" (Jones et al. 2008) or the number of false positives divided by the total number of positives. This could be the number of units flagged as poor performers when they're not in fact poor divided by the total number of units rightly or wrongly flagged as poor. For example, if the null hypotheses of 100 hypothesis tests were experimentally rejected and a maximum FDR level (known as the q-value) for these tests was 0.10, then fewer than 10 of these rejections would be expected to be false positives. The procedure for deciding which of your statistical tests is significant was developed in detail by Benjamini and Hochberg (1995) and involves choosing the FDR in advance and then ranking the p values for each test. The smallest p value has a rank of $i = 1$, the next smallest $i = 2$, and so on. Compare each individual p value with its Benjamini–Hochberg critical value, $(i/m)Q$, where i is the rank, m is the total number of tests and Q is the false discovery rate you choose. The largest p value that has $p < (i/m)Q$ is deemed significant, as are all of the p values smaller than it. If you use this procedure with an FDR greater than 0.05, it is quite possible for individual tests to be significant even though their p value exceeds 0.05. This is not wrong and reminds us that the conventional cutoff of $p < 0.05$ for statistical significance is not cast in stone. Sometimes you will see a "Benjamini–Hochberg adjusted p value", though not everyone agrees with their use as they don't really estimate the probability of anything (McDonald 2014).

The Bonferroni correction and Benjamini–Hochberg procedure assume that the individual tests are independent of each other. This is true when you compare unit A with unit B and then compare unit C with unit D. It's not true when you compare unit A with unit B and then compare unit A with unit C. There exist some more complicated procedures for dealing with this non-independence which are more relevant to analysing data from microarrays, for instance, and so will not be considered here. With benchmarking, we often compare each of our set of units against some fixed comparator like the national average, giving us independent tests. For further discussion of the multiple testing problem regarding comparisons of healthcare providers and focus on the false discovery rate, see Jones et al. (2008).

In addition to having multiple units to compare, monitoring involves multiple time points at which to make these comparisons. This increases the potential for false alarms. The problem of multiple time points is explicitly recognised in the thresholds of SPC charts such as CUSUMs. Thresholds can be raised in recognition of the length of monitoring as well as in recognition of the number of units being monitored; in Marshall et al. (2004a), we discuss the assumptions underlying the chart performance measures and offer some preliminary calculations of the FDR in this situation. Thresholds can also be tailored to take into account the resources available by the monitoring organisation to investigate any resulting alarms (Bottle and Aylin 2011b). Suppose there are 100 units that each treat 1000 patients per year. Setting a false alarm rate of 0.1% would be expected to result in 0.1% * 1000 * 100 = 100 false alarms, on top of however many true alarms are generated. If there's only the capacity to investigate, say, 10 alarms per year, then the FAR is too high. This simple back-of-the-envelope calculation assumes that each unit is only monitored on a single performance measure – having more measures will mean getting more false alarms, even with data of unimpeachable quality.

9.4.1 Multivariate Statistical Process Control Methods (∗)

With more than one indicator of interest, what we've covered thus far assumes that the indicators are independent. It's usual for each indicator to be monitored individually, which makes them easier to construct and interpret. It also means that each one has its own associated false alarm rate. The number of indicators can be reduced by producing a composite measure, as we saw in Chapter 8, but there still might be more than one composite of interest. When indicators are correlated with each other, joint (bivariate or multivariate) monitoring is more efficient and can lead to lower false alarm rates. A unit might perform just within the acceptable range for each individual indicator but, when considered as a whole using the whole set of indicators, could be quite different from other units. This multivariate approach is common in industry, where overseers of a manufacturing process necessarily need to keep tabs on perhaps hundreds of parts, but has been slow to make the transition to healthcare. An early example is by Steiner et al. (1999), who used CUSUMs to monitor death after neonatal surgery as one outcome measure and the need to reinstitute cardiopulmonary bypass after a trial period of weaning as the other. They defined simultaneous CUSUM statistics based on two univariate CUSUMs. The resulting "SCUSUM" chart was found to be particularly good at detecting simultaneous moderate increases or decreases in both measures.

The most familiar multivariate procedure is the Hotelling T^2 control chart for monitoring means. Developed in the 1940s, it's a direct analogue of the univariate Shewhart chart described earlier. It assumes that the measures of interest are normally distributed. Two versions exist: one for individual-level data and another for aggregate or group-level data. For the latter, we need to know the standard deviations of each measure and the covariance. The control limits for the chart involve a chi-squared statistic. If the mean of either or of both measures rises or falls sufficiently from its respective mean, then the control limit will be exceeded.

Like Shewhart charts, Hotelling T^2 charts ignore past information and only consider each time point by itself. To detect trends, CUSUMs or EWMAs are recommended. Lowry et al. (1992) developed a multivariate EWMA (MEWMA) control chart. MEWMA and EWMA are robust to the assumption of normality if used with measures that can reasonably be assumed to follow a normal distribution. Small values of their parameter lambda are proposed for general purpose monitoring. See Testik and Borror (2004) for a full treatment of how to design a MEWMA monitoring scheme.

Just as one can monitor the variability of a single indicator using the s^2 chart, there are ways to monitor the variability of a set of indicators, given by the covariance matrix. This can be done in a couple of ways. See Alt (1985) for an introduction to the problem and technical details.

These multivariate control charts are fairly effective if the number of measures being monitored is modest, for example, at most 10. As the number of measures n increases, the speed at which unusual performance is detected falls because the change to be detected gets "diluted" in the n-dimensional space of the measures. There are two common alternative ways of approaching this, particularly in situations when much of the variability or unusual performance occurs unevenly across the measures. Of these, the most common is the method of principal components, which converts the original set of measures into a much smaller set of new variables that nonetheless accounts for as much of the variation in the original as possible. The second method is that of partial least squares, though there's been little comparative work done to ascertain the relative merits of this method compared with other approaches.

One common difficulty with any multivariate control chart is how to interpret an alarm: which of the measures (or which subset of them) is responsible? This question is not always easy to answer. The standard practice is to plot univariate charts on the individual measures, but this isn't always successful for the same reasons that led to the use of the multivariate chart. One suggested way to reduce the number of false alarms is to introduce a Bonferroni-style correlation to the control limits. Another option is to monitor principal components, particularly if they can be readily interpreted in terms of the original variables, which is not always the case. Runger et al. (1996) recommend decomposing the T^2 statistic into components that reflect the contribution of each individual measure and then focusing on the components that contribute the most to the statistic.

9.4.2 Further Reading on SPC

There is a vast literature on SPC. In the health sphere, it has its uses in public health, post-marketing adverse event surveillance and in environmental monitoring as well as in our area of healthcare performance. The preceding section serves as an introduction to the methods that we think are most relevant to performance monitoring at the moment. Beyond some primers in medical journals aimed at healthcare staff, much of the literature is technical. Those readers wanting a rigorous overview could try the book by Montgomery (2012), now in its seventh edition; its Chapter 11 covers multivariate methods. For lots of worked examples of run and control charts in healthcare with an emphasis on quality improvement, see, for instance, the very practical book by Carey (2003).

9.5 Ways of Assessing Variation between Units

Before we discuss how much variation is "acceptable", we first need to consider the different approaches to estimating how much variation exists between units. The simplest measure is just the range. Related to the range is the ratio of the maximum to the minimum: for example, we say there is twofold variation between units. This is intuitive but is the least robust – it's very sensitive to unusually high or low values, which can occur just because of sampling error. More robust is to use other percentiles such as the interquartile range (IQR), though the IQR of course only covers half the distribution. The 5th and 95th percentiles would be better, though any choice would be arbitrary. In Chapter 7 in the context of multilevel models for binary outcomes, we met the 80% interval odds ratio (IOR 80%), defined as the interval around the median of the distribution that comprises 80% of the OR values. Other measures include the familiar standard deviation, useful only for normally distributed indicators, and the less intuitive coefficient of variation (SD divided by the mean).

Those approaches all ignore sampling variation, that is, they simply consider the difference in the estimated rates or other quantities. When units vary appreciably in size, it can be more useful to look at the amount of variation beyond just random noise. Some people therefore distinguish between random and systematic variation using the difference between what's observed and what's expected. Two such popular measures are the systematic component variation (SCV) and the empirical Bayes estimate (EB). Values different from zero mean there are systematic differences across units. In other words, high SCV or EB values suggest that a significant part of the observed variation could not be deemed merely random. A comparison of the two in the 1990s of small area variations in hospitalisation rates for a range of patient groups found that the SCV overestimated the median amount of systematic variation compared with the EB and that EB estimates were better able than SCV ones to predict future area-specific hospitalisation rates (Shwartz et al. 1994). EB tends to be more robust to small numbers and perform well under the null hypothesis, according to a more recent analysis comparing these two and other measures for their usefulness also in small area variation in healthcare utilisation (rates of surgery) (Ibáñez et al. 2009). Interestingly, they found that some statistics that they investigated perform differently under the null hypothesis depending on the rate of the surgical procedure. The EB and SCV didn't have this problem and seem best for general use. The authors concluded that the EB edges it and should "become an essential part of small area variation analysis studies". The same message applies to performance monitoring, but we suspect that the range and IQR are still the most popular.

9.6 How Much Variation Is "Acceptable"?

This question is particularly relevant to those definitions of outliers that make reference to between-unit variation. If the target is zero infections so that an outlier is defined to be any unit with at least one infection, then in theory it doesn't matter if no unit being monitored achieves zero infections, as anything other than zero is unacceptable. In practice, such an occurrence would likely lead to a rethink of the target. For definitions based on something such as whether a unit is "worse" than the average, however, there is an inherent concept of "acceptable" variation around that average. This is worth considering in more detail.

The issue is perhaps most clearly demonstrated in funnel plots because the control limits define the limits of acceptable performance. The standard way to draw them makes reference to some underlying distribution such as the binomial when plotting the proportion of adherence to protocols or the Poisson when plotting counts of infections or SMRs. This assumes that the measure that we're plotting at unit level follows the selected distribution (which it might not do) and so will exhibit random variation around the national average. Suppose that we're interested in the proportion of eligible patients receiving beta blockers at discharge and that it's not realistic or desirable to set the target at 100%. We therefore have to rely on reference to the national average proportion. If 1000 units all had the same underlying performance and, due to sampling error alone, varied around the average with nothing more than binomial error, then 1 unit on average would be expected to lie above the upper 99.8% control limit and 1 unit below the lower one. These unfortunate units and any other units lying outside the limits due to unusual prescribing rates – high or low – would be declared outliers. If we see more than 2 out of 1000 outside the limits, then our measure is said to be overdispersed: there's more variation than predicted from simply the binomial distribution. With just a few units outside the control limits, we could go and investigate them all. But suppose we have lots of units outside the limits. What should we do?

The plot can't tell us the cause of this overdispersion, so we can't distinguish between genuine variation in quality, where there are lots of genuinely good and poor prescribers, and artefact, such as the way that care is organised at those units (which may be different from at other units for good reason) and data errors. Artefactual errors afflict outcome measures much more than process measures, as imperfections in risk adjustment and the recording of administrative data come into play, as we have already seen in previous chapters. All these sources of "noise" can add up and cause the overdispersion, though at some units they will also cancel themselves out. There is debate over the most appropriate course of action, and it differs depending on your perspective. With overdispersed indicators there are three options:

1. Discard the indicator entirely.
2. Leave the control limits and hence outlier definitions alone.
3. Adjust for the overdispersion by widening the control limits.

As we saw in Chapter 7, random effects models produce "shrunken" SMRs so that small units are made to move towards the overall average. For example, small units with no observed deaths will have their estimated SMR of zero that they'd have from a fixed effects model modified into something much nearer 100 and probably within the control limits. Overdispersion among the larger units, however, would remain.

Proponents of the first option would argue that much of the overdispersion is data noise of whatever cause and that the resulting follow-up investigation of outliers would be plagued by false alarms (i.e., would have a low positive predictive value [PPV]). This can be a reasonable position to take, though politics and other pressures may trump it. For developers of the indicator, it's at least a prompt to undertake further analysis if possible. This could include correlations with other indicators, subgroup/stratified analysis by patient group to look for a more homogeneous denominator and a revisiting of any risk-adjustment model used.

The last two options are probably the most likely to be considered. If the intention is simply to show how much variation there is, then no adjustment for overdispersion should be made. For monitoring purposes, however, it should be considered, especially under the following conditions. One is when there are only limited resources to investigate outliers and there are too many apparent outliers for the system to handle. Another is when there is good evidence that most of the extra variation is due not to quality of care (which is what we're looking for of course) but instead due to data or other artefacts. In practice, such evidence is likely to be lacking. In addition, due to the nature of summing multiple errors and sources of noise, widening the control limits isn't guaranteed to increase your chances of picking the genuine outliers among the false ones, that is, of raising the PPV.

If you decide that at least some of the overdispersion is due to data issues, then the amount of variation considered acceptable is now greater than mere random variation. It now includes that due to data noise. Of course, we'd always like perfect data, but they don't exist. For the third option, we can allow extra permissible variation in an additive or a multiplicative way. The choice of estimation method affects the shape of the funnel plot's control limits. See Spiegelhalter (2005) for a comparison of the two and relevant formulae. The amount of extra variation beyond that expected due to chance can be estimated using random effects or quasi-likelihood modelling, though neither can distinguish between data noise and real variations in quality of care – there's no guarantee of retaining the baby at the expense of the bath water. We make the case against adjusting for overdispersion for performance monitoring in Marshall et al. (2004a, p. 556).

Counting funnel plot outliers is not the only way of estimating how much variation there is between units. Many studies describe the variation between units, especially via the range of values. However, when the units are small, a surprisingly large amount of this apparent variation is just random error. Diehr et al. (1990) noted this problem and that there were no good statistical methods to decide whether the variation between units is greater than that expected by chance. Using simulation and real data examples, they concluded that what seemed at first glance to be large variation between units often wasn't in fact so large. They also made the point that the four measures of variation that they considered (ratio of maximum to minimum, coefficient of variation, systematic coefficient of variation and chi-squared statistic) differed in their relative usefulness depending on the situation. They couldn't recommend one for all cases. A particular difficulty was caused by performance measures that could count the same patient more than once, such as falls or readmissions (when not expressed as the proportion of patients having one or more). They concluded that simulation would often be required.

Another method, specifically for SMRs and developed in an epidemiological context, takes the assumption of random Poisson variation for the observed rate ratios and assumes that the set of *true* rate ratios follow a gamma distribution. It is these true rate ratios (or true SMRs) that we are interested in, rather than merely the estimates thereof that we actually get from our data. The authors provide a simple test of homogeneity of SMRs (Martuzzi and Hills 1995). A low p value for this test would suggest that the true SMRs show more than just random variation, that is, some are genuine outliers.

The much-mentioned multilevel models yield an SD of the between-unit variation in outcomes when units are fitted as random effects (see Chapter 7). This SD between units is not much use by itself. Better is to assess the proportion of variation that is attributable to the unit level by calculating the intraclass correlation coefficient (ICC). This is the ratio of the variance between units to the total variance and so shows how much of the variation in outcomes is being driven by units and how much is being driven by patient factors. The larger the ICC, the more important the units are compared with patient factors. For logistic regression, however, there's a problem when trying to calculate the ICC, as the patient-level variance is on the probability scale and the hospital-level variance is on the logistic scale. A technical way round this (the linear threshold approximation) was proposed, in which the patient-level variance is given as 3.29 (Snijders and Bosker 1999). There are other problems, however, with the ICC for logistic regression that limit its use, such as its dependence on the outcome prevalence. After all, like the R^2, it's a measure devised for linear regression that people have tried to adapt for the non-linear case. Instead of either the SD between units or the ICC, Larsen and Merlo (1995) suggested calculating the median odds ratio (MOR) between the unit at higher risk and the unit at lower risk when randomly picking out two units. The median odds ratio is a measure of variation between the mortality rates of different units that is not explained by the modelled risk factors. If the MOR is 1, there is no variation between units. What none of these measures tell us, though, is how big is "unacceptable".

9.7 Impact on Outlier Status of Using Fixed versus Random Effects to Derive SMRs

We now consider a question that has attracted more attention recently: what is the impact of using fixed and random effects on the number of outliers? As described in Chapter 7, SMRs can be derived from fixed effects models, in which each unit is assumed to have its own performance distribution, or random effects models, in which each all units are assumed to belong to a single performance distribution. There are both simulation and empirical studies on this question. The latter often found poor agreement between the two approaches on which units were labelled outliers: see Austin et al. (2003) and references therein. With larger numbers of patients per hospital, however, we found much better agreement (Alexandrescu et al. 2014). These studies have the key problem of a lack of a gold standard – empirical studies inevitably don't know which hospitals are genuine outliers so can only compare methods in terms of the number of units given the label.

Austin et al. (2003) got round this limitation using simulation. They generated 110 synthetic units and for post-AMI 30-day mortality compared simple O/E ratios from logistic regression, as commonly used in hospital report cards, with empirical Bayes odds ratios, now used by the National Surgical Quality Improvement Program as described in Chapter 7. Risk adjustment was assumed to be perfect. They stratified analyses by patient risk, running simulations for average-risk patients and for high-risk ones. Their gold standard definition of a high-mortality outlier was one at which the patient had 30% or higher odds of death compared with at an average unit. They further considered two different distributions of hospital performance: a unimodal one where all units had the same underlying mortality rate and a trimodal one where there was a mixture of low-, average- and high-mortality units. Briefly, for the unimodal case, the random effects models had appreciably higher PPV, especially for high-risk patients. On the other hand, the fixed effects models had much greater sensitivity, especially for average- or low-risk patients. For the perhaps more realistic trimodal case, however, the fixed effects models had slightly better both PPV and specificity than the random effects ones. It's worth noting that the PPV was highest in the unimodal scenario for average- or high-risk patients and lowest in the trimodal scenario when the proportion of high outliers was low. Another point to note is that the lower sensitivity of the random effects models was likely due to the shrinkage of their SMRs towards the mean. Even when the authors ran simulations for large hospitals (>250 patients), the random effects models picked up fewer genuine outliers. Their results contrasted with those from the earlier simulations by Hofer and Hayward (1996), who assumed a bimodal distribution for hospital performance (a mix of average and poor units) and varied the proportion of deaths due to poor care rather than look at mortality rates for specific patient profiles as Austin et al. (2003) did.

Austin et al. (2003) concluded that the unimodal case represented a classic trade-off between sensitivity and specificity. If you want to catch more poor performers, then use fixed effects models. If you want to minimise the false alarms that you have to investigate, then use random effects models. In the trimodal case, the fixed effects models performed a bit better on both counts, though the authors noted that the trimodal scenarios are settings in which the use of conventional random effects models may be inappropriate. The reason is the assumption called exchangeability that such models make, where all units are assumed to have the same a priori mean, that is, they have roughly similar performance. Scenarios in which there are three distinct distributions of hospital-level death rates – low, average and high as in their trimodal scenario – clearly violate this. Another common scenario that violates the assumption is when there is a relation between volume and outcome, as often demonstrated in surgery. Fedeli et al. (2007) looked at breast-conserving surgery rates and suggested that the volume–outcome relation reduced the sensitivity of the random effects models.

Part of the study by Kalbfleisch and Wolfe (2013) used simulation with a linear outcome to assess the effects of two assumptions made by random effects models: (1) that the unit effects are normally distributed and (2) that the unit effects are independent of casemix. The latter assumption may plausibly be violated if sicker patients are referred to better units (those with better outcomes) or if they live near worse-quality units. Fixed effects models are not affected by these problems. To get round this, one could take a two-stage approach. In the first stage, a fixed effects model is fitted that includes the covariates and a risk score derived. In the second stage, this risk score is entered into a random effects model as an offset. However, Kalbfleisch and Wolfe's results suggest that fixed effects models are still better at detecting units with extreme performance even after such adjustment.

Is there an easy answer to the question of whether we should monitor outcomes rates or ratios from fixed or from random effects models? Not really, as is attested by the differences in opinion. We in the United Kingdom have generally come down on the side of fixed effects, whereas in the United States several authorities prefer random effects, albeit of differing specifications as described in Chapter 7.

9.8 How Reliably Can We Detect Poor Performance?

We end this chapter by following on from the previous section with some comments on the key question. As usual, the exact formulation of this question suggested by the heading to this section depends on the main goal of the monitoring, given that no screening test is perfect. Depending on such non-statistical factors such as culture and resources, high sensitivity or high specificity might be more important. We have just seen how, in general, fixed effects models favour the former and random effects ones the latter, though the studies by Austin and by Kalbfleisch and Wolfe (2013) show that things are more nuanced than this.

Several U.S. studies have assessed the reliability (signal-to-noise ratio) of physician and hospital report cards. Adams (2009) used claims data to classify physicians in different specialties in terms of their cost performance, simplified as belonging to one of just two tiers. Reliability (otherwise known as the intraclass correlation coefficient described earlier) ranges from 0 to 1: 0 means that all the variation in cost-profile scores is the result of measurement error (intra-unit variation), and 1 means that all the variation is the result of real differences in performance (inter-unit variation). Overall, 59% of physicians had cost-profile scores with reliabilities of less than 0.70, a commonly used threshold for adequate reliability, and 22% of physicians would be assigned to the wrong tier. Vascular surgery, with the fewest physicians and the fewest mean episodes per physician, rated particularly poorly. Increasing the sample size by adding more years of data didn't make a dramatic difference. The authors recommend that users of physician cost profiles directly assess reliability instead of relying on minimum sample sizes. Users would have to decide how to classify the many physicians who didn't meet the 0.70 minimum reliability threshold. One option is to report that there wasn't enough information to classify them. Sample size is not the only determinant of reliability, though. As the authors point out, an aim of quality improvement efforts is to reduce variation between units. Reducing this variation would actually *lower* the reliability: if all units had identical performance, the coefficient would be zero. Normand et al. (2007) list four determinants of how many cases are needed to accurately classify units into tiers:

1. The average level of the performance measure
2. Its variability across units
3. The number of patients within a given unit who are eligible for the measure
4. The desired level of accuracy

Using simulation based on real data from two U.S. states, they estimated the minimum number of patients needed for each process of care measure to classify a unit as "excellent" (>80% or >90% of patients receiving the right therapy). Depending on the measure and performance variation between hospitals, the proportion of hospitals with enough patients to spot good performance ranged from 0% to 98%. The finding that sample sizes are too low for some measures and patient groups, particularly for mortality for some types of surgery, had been noted before (Dimick et al. 2004). Other studies have estimated the reliability for other settings. A much-cited study by Hofer et al. (1999) came out heavily against primary care physician report cards for diabetes. More encouragingly, Krell et al. (2014) found that overall complications and reoperation, but not mortality, had adequate reliability for bariatric surgery at the bigger centres. Kao et al. (2011) found that surgical site infections following colon resection had high reliability with as few as 94 cases.

All those kinds of studies used statistical methods to estimate the signal-to-noise ratio for a given setting and indicators. They're useful, but the most convincing evidence of whether a given indicator can consistently and reliably detect real outliers (true positives) is much more limited as it requires reference to a gold standard marker of performance. Our old friend, aggregate hospital mortality, has long been criticised partly because such evidence was lacking. Some finally arrived in the shape of the Keogh (2013) report into 14 hospitals flagged as having significantly high overall mortality, as measured by either the hospital standardised mortality rate (HSMR) or the summary hospital-level mortality indicator (SHMI). Inspectors visited the 14 sites, found problems at all of them, and put 11 on "special measures" due to the serious nature of the problems. While the HSMR and SHMI did not always identify the same hospitals as being high, 11 out of 14 represent a respectable PPV of about 80%. Of course, the flip side is missing from this: we don't know how many of those hospitals that were *not* flagged as outliers had problems of similar severity during the same time period. This also presupposes that the inspector team's assessment is truly golden; the national regulator in England, the Care Quality Commission, underwent a change of leadership in 2013 due to well-publicised difficulties with its inspection regime. However, what constitutes good and bad quality of care will in any case continue to evolve.

9.9 Some Resources for Quality Improvement Methods

Performance monitoring and outlier detection obviously comprise one part of the monitoring cycle. While this is not meant to be a book on quality improvement methodology, we close this chapter with a few popular resources for the other vital part of the cycle: improving performance.

The U.S. Institute for Healthcare Improvement (IHI) works with institutions big and small across the world to apply improvement science and trace their methods back to WE Deming in the early to mid-twentieth century. Their method is called the Model for Improvement, developed by Associates for Process Improvement (www.apiweb.org). It asks the following three key questions of any quality improvement project. What are we trying to achieve? How will we know that a change is an improvement? What changes can we make that will result in an improvement? Key to the method is the series of plan–do–study–act cycles. Their website has a host of white papers, tools, case studies and guides. Several books came out of IHI's efforts, and those by Berwick et al. (2002) and Langley et al. (1996) remain some of the best descriptions of their approach, describing what does and what doesn't work in practice. The latter is particularly popular and has a new (2009) edition that also shows how to accelerate improvement by spreading changes across multiple sites.

From the United Kingdom, the book by Boaden et al. (2008) is a reflective review of the main concepts and tools for quality improvement. It originated as part of the evaluation of the Six Sigma training program run by the National Health Service (NHS) Modernisation Agency. Its focus is therefore on the approaches and methods that industry has used successfully over the past 50 years and what they mean for healthcare. For example, the authors conclude that elements such as teamwork, commitment to improvement and leadership are more important than the choice of particular improvement method (though a method is needed!). The book is available free for download. The NHS Institute offers 15 Improvement Leader guides, available online as a box set of pdfs that cover three themes: general improvement skills; process and systems thinking; and personal and organisational development. North of the border, NHS Scotland's Quality Improvement Hub website provides a range of resources, including but certainly not limited to the Model for Improvement and IHI's Global Trigger Tool.

Specifically for patient safety, see Charles Vincent's book that we referenced earlier (Vincent 2010) and Patient Safety First's how-to guide for measurement for improvement (again, this is focused on the Model for Improvement). WHO's Patient Safety Programme covers elements such as education and training, how to implement change and patient engagement and is found at http://www.who.int/patientsafety/implementation/en/. It includes its famous WHO Surgical Safety Checklist; other patient care checklists are being developed.

10

Making Comparisons across National Borders

Aims of This Chapter

- To set out the rationale for multi-country comparisons
- To give some examples of cross-national databases of patient-level records
- To illustrate the challenges involved in such comparisons, focusing on the statistical and interpretational ones

Most comparisons of performance are made of units within the same region or the same country. Compared with multinational sources, local data are more likely to be available and consistently defined and collected, and the units are subject to the same healthcare system, which helps with interpretation of any performance variations. Nonetheless, there are good reasons to look further afield. Most governments are eager to get more for their healthcare dollar, and it's natural to look abroad for potential solutions to the common problems of rising costs, health market failures, poor quality and safety and inequalities. A country or indeed any healthcare unit that claims to be a world leader ought to have evidence for those claims, and such evidence can't come from a merely national information source.

There are two ways to compare units across national borders: to calculate indicators using each country's sources separately and to pool patient records first before such calculations. Several organisations such as the World Health Organisation (WHO), Commonwealth Fund and the Organisation for Economic Cooperation and Development (OECD) regularly compare health and healthcare systems on spending and performance using the first option. The WHO supports countries "to develop, implement and monitor solid national health plans… and to strengthen health information systems and evidence-based policy-making", and we saw an example of how they compare national health systems in Chapter 4. There are several reasons why units in different countries join forces to produce an international patient-level database that allows comparisons of performance, which is the second option. In many ways, such a database is just an extension of a national one that's used for benchmarking and mutual learning, but it has other advantages. It provides greater statistical power, which is particularly important for small countries and/or rare diseases. Second, it may allow the assessment of the impact of healthcare system on treatment and outcomes. Cross-country databases can have different origins: multicentre randomised controlled trials (RCTs), national registries and administrative databases. For performance monitoring, a collection of data from traditional RCTs will be of limited use for several reasons. First, the trials have predetermined start and end dates that fixes the data collection period in time, which limits the currency of the performance estimates. No prospective monitoring is possible. Second, typically tight exclusion criteria for trials limit the generalisability of the results to enrolled patients, who are on average younger and fitter than the population they are meant to represent (though we consider more inclusive "pragmatic trials" in Chapter 12). Third, the aim of trials is to compare a typically newer treatment against an older one rather than to compare performance between units, so they won't necessarily involve the collection of relevant data. A comparison of units from such a database is inevitably a comparison of units who know they are taking part in a trial, which will have its own effects on the results of such a comparison.

National clinical registries and administrative databases were discussed in Chapter 5 in terms of their features and their pros and cons in general. Before we describe the main hurdles that such projects have to overcome and some options for tackling the statistical problems, we will give a few diverse examples.

10.1 Examples of Multinational Patient-Level Databases

- Global Registry of Acute Events (GRACE)
- European Cystic Fibrosis Society Patient Registry
- Global Comparators Project
- European Collaboration for Health Optimization (ECHO)

At the time of this writing, the Global Registry of Acute Events covers over 200 hospitals in 28 countries and is the largest multinational registry that covers the full spectrum of acute coronary syndrome (ACS). One of its main objectives is to describe diagnostic and treatment strategies and both in-hospital and post-discharge outcomes. Its aims are to improve care and inform future research. Rather than try to capture all cases, which many disease registries try to do, it asks for the first 10–20 consecutive cases per centre per month. It's been used in a number of journal publications, such as an analysis of variations in discharge medication and outcomes by country (Fox et al. 2002). Data quality is maintained by regular audit. Participants receive confidential quarterly reports with benchmarking of their patients' characteristics, presentation, management and outcomes against themselves, their cluster (e.g., North America) and the world. It's supported by a grant from the pharmaceutical giant Sanofi.

The European Cystic Fibrosis Society Patient Registry has collected data on around 30,000 patients from >20 countries. It grew out of a database set up to monitor a clinical trial with funding from pharma that ended in 2002 and now takes data from individual cystic fibrosis (CF) centres and national CF registries. Specialists from the European CF Society decided to set up an independent registry with comparable aims to those of GRACE. An analysis in which patients were stratified by lung function, age and disease severity revealed which practice patterns were associated with better outcomes (Padman et al. 2007). Due to its piecemeal creation from several existing registries, each with their own ways of doing things, it had some extra challenges around case definition and data collection. Even if the individual registries provided specifications on data formats and coding, they didn't always follow them.

The Global Comparators project, with which we (AB and PA) are both involved, is facilitated by the UK-based company Dr. Foster and began in 2010. It brings together leading hospitals from a growing number of countries for mutual learning. Specialty-specific groups of clinicians meet regularly by phone and face to face at biennial conferences to agree on research and quality improvement projects and to monitor progress. Rather than use registries, the project takes the more pragmatic option of combining existing administrative data supplied by each participating hospital into a project database. This is then interrogated by analysts and participants; an interactive online tool also aids analysis of the database and acts as the starting point for discussions around apparent differences in outcome within and between countries. Much attention at the start was inevitably given to the difficulties in adapting the records submitted by each hospital to best enable like-for-like comparisons of, for instance, mortality following colorectal surgery for cancer. An illustration of this is given later in this chapter for stroke mortality as a worked example.

Inspired by the U.S. Dartmouth Atlas of Health Care, several European countries developed their own national atlases of medical practice variation, and indeed the OECD published a monograph with variation studies in 13 countries (OECD 2014). The European Collaboration for Health Optimization project grew out of Spain's Atlas (www.atlasvpm.org) and brings together national hospital databases of several European countries to show variations in performance both within and between countries. It uses administrative data on hospital discharges, healthcare supply, population and socioeconomic measures. The February 2015 supplement to the *European Journal of Public Health* covers several studies from the project, and an editorial describes the lessons of the ECHO as being both practical and methodological. The project's methodology handbook (ECHO 2014) sets out in detail each country's data sources and how the team overcame the technical data challenges such as differing coding systems and transforming variables into a standard format, some of which we also encountered in the global comparators (GC) project.

10.2 Challenges

Many of the challenges that an international database faces are the same as those faced by a national or even regional one, such as funding, logistics, information governance and data item standardisation. To these one can add language and potentially also cultural barriers that can affect many aspects, from case finding and data entry through to presentation of results. There can be culture-specific sensibilities regarding certain data items, for example, lifestyle elements, mental health conditions and sexuality, or even the outcome measures. "Soft" outcomes such as patient satisfaction scores are notably affected by social, ethnic or cultural group and will need validating in each target population before use. Table 10.1 lists the principal challenges for an international database. For a review of how one registry dealt with these, see Viviani et al. (2014) on the European Cystic Fibrosis Society Patient Registry. For a guide on how to use registries in particular, from the planning phase through to evaluating the registry itself, see the Agency for Healthcare Research and Quality (AHRQ) report (Gliklich and Dreyer 2010). We concentrate on the statistical and interpretational ones.

The main challenges that we now consider are

- Combining administrative databases from different countries, including risk adjustment
- Modelling country effects
- Dealing with partial data collection systems and interpreting differences in country-level performance measures

These have already been discussed more generally in earlier chapters, so perhaps the best way to illustrate how they need careful thought when comparing units across national boundaries is with an example.

TABLE 10.1

Main Challenges for Maintaining an International Database for Benchmarking Outcomes

Challenge	Comments
Setting the objectives	The biggest questions are what to measure, how and why. It is best to agree with patients to help get their consent to share their data. The objectives will guide what data items are collected.
Defining cases	This requires generally agreed diagnostic criteria, which may change over time due to advances in knowledge. Confirmed cases only (generally the best option) or those with a suggestive set of symptoms or signs? Beware of the effect of early adopters on outcomes when looking at use of new methods, e.g., surgical technique.
Defining the variables to be captured	You need to decide which elements of the disease to capture. Overenthusiasm can be costly in terms of data quality. Strict definitions used in trials can differ from what's used in the clinic, so compromise – but data must be useful to clinicians or the registry will fail. Compromise is also often needed to accommodate differences between countries' existing definitions and coding systems.
Data management	Preferably automate the system, from data collection and checking to presentation of results. Best practices include making derived variables centrally to minimise error.
Data quality controls	This is clearly crucial and also harder if database is overambitious. See Section 5.1 on how to assess data quality.
Handling of missing data	Many registries suffer greatly from missing data: less so for demographics and more so for clinical follow-up. Mechanisms causing the missing values need to be understood: e.g., if combining existing registries, some countries' registries won't collect certain items, so those fields will always be missing for that country; follow-up losses due to adverse events or poor adherence to treatment can bias comparisons.
Maintaining confidentiality	The data sender must be legally allowed to send the data, and the central registry must be legally allowed to process the data. The registry needs to keep up to date with changing data legislation.
Data use arrangements	Objectives of a disease registry also allow use of data for identifying patients eligible for clinical trials, though patients must be contacted by their care provider. Use by researchers is essential to maximise value, though access will need to be managed, e.g., by a scientific committee.
Dissemination of results	Obviously this needs to allow for different languages and types of stakeholder. Benchmarking requires agreed benchmark(s) and risk adjustment if necessary.

10.2.1 Worked Example of Combining Administrative Databases from Multiple Countries: Stroke Mortality

As outlined earlier, the Global Comparators project incorporates administrative data from multiple countries into a single database with which participants compare their unit's outcomes with those of others. The default benchmark is the overall average for all participants combined. Suppose one wishes to compare mortality following admission for stroke. For the database as a whole, we on the project committee first had to distinguish between an inpatient admission and a stay in an observation or assessment unit. Not all countries had a flag to indicate a stay in such a unit (and it wasn't always reliable) so we had to make use of length of stay, procedure and destination on discharge information. Unplanned same-day discharges not ending in death or transfer and without any major procedures were assumed to be observation or assessment ward stays.

Next, to identify admissions for stroke and other conditions, International Statistical Classification of Diseases and Related Health Problems (ICD)9 and ICD10 codes had to be grouped into usable diagnoses. This was first done using AHRQ's Clinical Classification Software (CCS). It allocates the primary diagnosis code to one of 259 clinically meaningful groups for health services research and is much used internationally. Participants used either ICD9 or ICD10 for diagnoses, and there exist look-ups to map each to CCS groups. Likewise, procedure codes had to be grouped into usable procedures, which presented a much greater problem due to the larger number of different procedure coding systems in use among project participants and the lack of the equivalent of AHRQ's CCS. We drew up an initial grouping based on our English groups, reviewed the surgical literature on non-UK administrative databases and consulted with coding experts in each country before sending out to review by participants. The project has yet to face the problem of country-specific definitions of the "primary diagnosis" in administrative data. The main reason for admission is not necessarily the same as the main problem treated or the problem that uses most resources. Other fields were not always defined in the same way, but categories could be combined. An example is urgency of admission, which in some countries is essentially just planned (elective) or unplanned (emergency) but in others can be planned, urgent or unplanned. We put urgent with unplanned. Our approach to these problems is described more fully elsewhere (Bottle et al. 2013a).

Regarding stroke specifically, the AHRQ CCS group corresponding to stroke is 101, "cerebrovascular disease". However, this includes subarachnoid haemorrhage, which the participating stroke physicians wanted to handle separately. The CCS group was therefore tweaked for the project to get the desired definition of stroke.

Perhaps the biggest challenge in this project was the outcome measure. All countries recorded in their databases deaths occurring in hospital. For England, we had linkage between the hospital records and the national death register to capture post-discharge deaths, albeit with a time lag, but we had to discard this information as it was unavailable for the other countries. The use of intermediate care and the length of stay in the acute hospital varied greatly by country. In England, a number of acute hospitals start the rehabilitation in that setting, but rehabilitation is not reliably identifiable in the data (in theory it's possible using the specialty field and secondary diagnosis fields via ICD10, but capture appears to be incomplete). In the United States, more patients are transferred out of the acute hospital and they are transferred earlier, so it isn't possible to find a short enough length of stay cutoff to make fair comparisons. In a pilot project, volunteer hospitals collected 30-day and 90-day modified Rankin scores (a score of 6 means the patient died) and National Institutes of Health Stroke Scale (NIHSS) on all ischaemic strokes for a month for linkage with the existing project database. This showed the superiority of using the modified Rankin scores as an outcome over in-hospital mortality (total and 7-day) (Rost et al. 2016).

Risk adjustment proceeded in the usual manner as in Chapter 6. Adjustment for comorbidity for the project is currently done using the Elixhauser set of conditions plus dementia, with weights tailored to stroke and the outcome of interest and derived from all stroke admissions in the project database. One assumption is that each comorbidity has the same relation with mortality in each country, which is probably untestable. The likelihood of a given comorbidity being recorded for a given patient during a given admission depends mainly on the coding guidelines in force at the time in that country and the quality of coding at that hospital. That in turn depends on various factors including the availability of unambiguous, legible medical notes as we saw in Chapter 5. The Netherlands has long had problems with its "LMR" administrative data. Recorded prevalences of every comorbidity in the risk model were higher in the United States than in the English hospitals, partly because of the more long-standing use of Diagnosis-Related Groups as a recording incentive there. While we found that comorbidity weights derived separately for each country showed similar patterns to those derived from all countries combined (Bottle et al. 2013b), the lower recorded prevalences in the Netherlands mean that their hospitals are "penalised" in the risk adjustment compared with England and the United States. Some Dutch hospitals were excluded from the risk-adjustment modelling because they had almost completely empty secondary diagnosis fields. It may be tempting to apply some kind of country-specific upcoding fudge factor to the Netherlands in particular, but the real level of comorbidity in Dutch patients is of course unknown. It may be possible to estimate this if there are studies comparing the same electronic database with the relevant casenotes (i.e., a chart review) for each country at similar time points. For stroke, the risk adjustment is in any case greatly hampered by the lack of a good severity marker in administrative data, which generally lack the validated NIHSS. However, ICD10 coding in the United States incorporated the NIHSS from October 2015. There will doubtless be some time lag before completeness levels make this useful.

Stroke represented an extreme case. Happily, for most of the other conditions that the project focused on, such as colorectal excision and hysterectomy, the data enable comparisons that are fair enough to avoid getting too bogged down in discussing artefactual explanations for differences in outcomes. The database is starting to generate peer-reviewed publications.

10.2.2 Clustering within Countries

The data exhibit clustering, as described in Chapter 3, though this time hospitals are clustered within countries as well as the patients being clustered within hospitals. The pros and cons and the mechanics of accounting for the clustering of patients within units were considered in Chapter 7, but what if anything should we do about the clustering of hospitals within countries? In theory, we should do something. If we use a multilevel model, should the hospitals and countries be considered as fixed or random effects? As their healthcare systems are quite different, one could argue that the countries are not "exchangeable" (an assumption of random effects models), so including them as random effects wouldn't be right. They are also few in number and of interest in their own right rather than representing some larger pool of countries. If the aim of the analysis is to compare countries, then the use of fixed effects would be best.

The aims of the project include exploring differences in outcomes between countries but also between hospitals. How should we adjust for country effects in this case? Again, the countries are not "exchangeable", so random effects would seem inappropriate. Adjusting for country by including them as fixed effects is also probably not a good idea. On the one hand, the part of the effect of hospitals on outcomes that in theory is due to its healthcare system (country) would be removed. On the other hand, what exactly would we be adjusting for? "Country" represents all manner of things from government policy on resource-setting, national clinical guidelines, casemix of admitted patients (which reflects not just levels of illness but also payment mechanisms, health-seeking behaviour and healthcare staff admission criteria, for instance) and coding levels as discussed earlier. For these reasons, the project has not accounted for the clustering of hospitals within countries in the risk-adjustment models.

10.2.3 Countries with Unusual Data or Apparent Performance

Depending on the size of the database, an individual country can have a noticeable influence on the overall mean and hence the benchmark if the overall mean is chosen for this role, as is often the case. This is analogous to an influential point in any data set: a common definition of an influential point is as an outlier (an unusually high or low value) that greatly affects the slope of the regression line. There can be different reasons for this. In the GC example, some Dutch hospitals were considered outliers due to their unusually low recorded comorbidity prevalences. The reason for their unusual data was artefact (poor data quality), and the hospitals were excluded from the risk modelling. Registries can suffer from selection biases that can result in artefactual differences in outcomes between countries or units: examples include case-finding difficulties in 10 of the 21 populations covered by the otherwise high-quality WHO MONICA database of cardiovascular events

(Asplund et al. 1995) and eligible patients who were missed out from a national acute coronary syndrome registry having higher mortality rates and poorer care than those enrolled (Ferreira-Gonzalez et al. 2009); see Chapter 5 for a discussion of the challenges of using each type of database.

If, on the other hand, a hospital or indeed a whole country can be considered an outlier for non-artefactual reasons, then they may be included – especially if it would be politically useful for the project – unless they are too influential on the benchmark. In a large multinational database, their influence may be absorbed by the rest just as one hospital's influence may be absorbed in a database covering a large number of hospitals.

10.3 Interpreting Apparent Differences in Performance between Countries

The previous two sections in this chapter have concerned only patient-level databases. In this section, however, we consider how to interpret the results of international comparisons of whatever data source.

The first potential explanation for apparent differences in performance between countries that must be considered is data artefact. We've already seen some examples of how international comparability can be threatened with both administrative and registry databases when cases or data items are not defined the same way or are not taken from the same type of database. These problems affect not just vital registration or morbidity data but also information on disease risk factors and healthcare staffing.

If the ideal set of data items is lacking, there can sometimes be reasonable ways of filling in the gaps or overwriting incorrect values; we covered imputation with individual-level data in Section 6.6. For instance, the annual WHO world health statistics reports cover nearly 200 countries, some of which lack death registration systems. Household surveys and censuses were used instead to estimate mortality rates and life expectancy. Some simpler recoding of the cause of death was possible. For example, if neonatal septicaemia (ICD10 P36) was instead assigned to A40 and A41 (septicaemia) without taking the age of the child into account, it could simply be substituted with P36 (WHO 2011). If the country was small but it did have "good" levels of death registration, defined as at least 85% coverage, three years of death data were combined to reduce the random variation. If death registration coverage was assessed as less than 85%, a procedure called cause-of-death modelling was used to adjust the proportions of deaths occurring for each of three major disease groups in the Global Burden of Disease study. The same approach was used to deal with miscoding for cardiovascular diseases, cancer and injury. It was used with those countries that lack cause of death data but that do have some estimates of overall death rates, not necessarily from vital registrations. With this procedure, data from a range of other countries, including some middle-income ones to help with representativeness, were used to provide the estimates of the proportions of total deaths that fell in each of the three major disease groups. Regression models were run for each age group and sex and included all-cause mortality plus income per capita in international dollars. Both variables were included in logged form to improve model fit (Salomon and Murray 2002). This approach is useful when forecasting the cause of death pattern for a country in three situations: where the most recent vital registrations data are for several years in the past, when estimating the pattern for areas in a country where only some areas have death data, and when estimating the pattern for countries that lack death data but where its neighbours in the region do have them (Mathers et al. 2006). Of course, these approaches only give estimates and are inferior to having the country's real data, but they're the best we can do.

If we take country-level measures at face value and put aside concerns over data comparability, then how should we interpret what are often striking apparent performance differences? Rather than leap to the conclusion that one healthcare system is better than another, Murray and Frenk (2000) argue that we should use information on a set of potential explanatory factors to investigate. The analysis can be done with a standard regression model. Before the modelling, some framework is necessary for thinking about which dimensions of the systems might influence performance. Factors could include elements of what type of system each country has (e.g., a national health system), what each system does (e.g., immunisations or intensive care units) and how it's paid for (e.g., tax based or social insurance based). The levels of healthcare spending and gross domestic product (GDP) could also go into such models.

Another point relevant to interpreting country-level figures is the timescale of the measurement of those figures. One consideration is the desire for regular and frequent information and evidence that any system changes have led to improvement. Another is the need to consider the cost of the exercise but also what the health system can realistically do in the time since the previous assessment. Many health sector reforms and institutional changes may take several years to have their full effect, which can be too long for health ministers to wait. If the timeframe is too short, health systems won't be held accountable for past mistakes – or rewarded for past successes.

10.4 Conclusion

In summary, most of the issues when combining existing databases or creating new ones that cross national borders are not statistical but financial, logistical and political. The issues that are primarily statistical in nature can also be found in studies involving multiple regions within the same country – but simply on a bigger scale. There are considerable variations in data infrastructure both between and within countries that affect which indicators can be used in international comparisons. Gaps can sometimes be mitigated, and regression modelling has been extensively used by the WHO to fill in such gaps in health and population registration data systems in some countries. Nonetheless, caution is still required when interpreting the results.

11

Presenting the Results to Stakeholders

Aims of This Chapter

- To describe the main ways of presenting healthcare performance data
- To show that it matters how the data are presented to staff and the public and to make some recommendations

There is little point in doing sophisticated analysis to produce validated measures of quality of care on expensively collected, reliable data if none of it sees the light of day – or if it's presented in a format that its intended users can't understand or act on. There are many possible ways of presenting data and a variety of potential audiences, though the latter can essentially be divided into two: staff and patients. Before providing examples of how performance data are shared with the public in different countries, it's important to consider whether feeding back such information to healthcare staff affects their behaviour and changes practices and outcomes. First, though, we'll briefly review the main ways of presenting performance data.

11.1 The Main Ways of Presenting Comparative Performance Data

Presentation formats used in comparative hospital quality reports may be categorised into three types (Gerteis et al. 2007; Damman et al. 2010):

1. Numerical formats that display numbers or percentages in tables
2. Graphical formats such as bar graphs
3. "Evaluative" formats that transform the information to a good/bad scale

With multiple indicators, one can either report on every single indicator value of each unit or aggregate information into a score or into categories,

TABLE 11.1

Fictitious Data for Average Appointment Waiting Times by Unit with Categories and Star Rating

Unit Name	Mean Waiting Time (Days)	Waiting Time Category	Star Rating
Ample Parking Hospital	34	Average	**
Concrete Yard Clinic	50	Worse than average	*
Farmyard County	29	Average	**
Metred Stay Hospital	18	Better than average	***
No Space Infirmary	38	Average	**
Taxi-Only Clinic	28	Average	**
Walk-In Centre, City General	35	Average	**

for example, grades instead of indicator values (see Chapter 8 on composites). These categories can have "evaluative cues", which are added to help the viewer interpret the information (Hibbard and Peters 2003). Examples of these cues include a star rating (each unit gets a number of stars as for consumer websites such as TripAdvisor), comparative terms like "better" or "worse", and traffic lights whose colours represent better-than-average, average or worse-than-average values. Table 11.1 shows fictitious unit-level data in tabular form with units allocated to one of three categories. Note that we've put the units into alphabetical order rather than league table (naïve ranking) order as a newspaper might employ.

Figure 11.1 extends the information to show a number of indicators. The values for each indicator have been replaced by a traffic light colour to show its position relative to the benchmark, which in this made-up example is the national average. Indicators have been selected from those used by the Scottish Government for its National Health Service (NHS) performance targets, known as HEAT targets.

A gray or a blank box can be used for missing data. Rather than using a simple block of colour, arrows can indicate trends. For example, NHS Greater Glasgow and Clyde in Scotland use green for meets or exceeds target, amber for

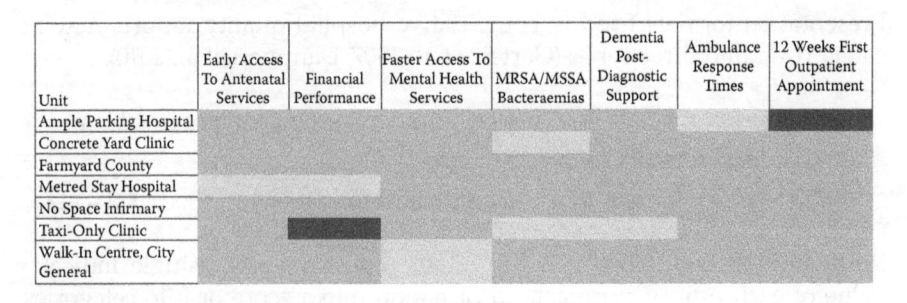

Unit	Early Access To Antenatal Services	Financial Performance	Faster Access To Mental Health Services	MRSA/MSSA Bacteraemias	Dementia Post-Diagnostic Support	Ambulance Response Times	12 Weeks First Outpatient Appointment
Ample Parking Hospital							
Concrete Yard Clinic							
Farmyard County							
Metred Stay Hospital							
No Space Infirmary							
Taxi-Only Clinic							
Walk-In Centre, City General							

FIGURE 11.1
Traffic light display of fictitious data on multiple indicators by unit.

within 10% of target, and red for greater than 10% from meeting target. A side arrow means that the performance is the same as for the same period last year, an upward arrow shows improvement and a downward arrow shows deterioration since last year. Of course, the closeness to the target that each colour represents and the definitions of improvement and deterioration can be set in many ways and in varying degrees of statistical sophistication. Sampling error might be ignored or it might be subject to the full blast of Bayesian shrinkage.

Figure 11.1 compares units but can be adapted to compare indicators for different patient groups at the same unit. For example, Guthrie et al. (2005) illustrate the relative merits of league tables and control charts and give an example of how measures relating to smoking, blood pressure, cholesterol, the appropriateness of the diagnosis and medications are shown for a given general practice for patients with coronary heart disease, stroke, diabetes, hypertension, asthma, chronic obstructive pulmonary disease (COPD) and also for all patients combined. The practice can be seen to do consistently well on diabetes, but it has issues around diagnosis concerning several patient groups. Whether the unit does "much better than the comparator" or "much worse than the comparator", for instance, depends on the indicator's value relative to the control limits on the control chart, but, as with our earlier examples, the reader doesn't actually see these underlying charts and hypothesis tests. This also means that indicators of any data type or scale can be displayed together.

As an alternative to rectangular shapes, a radial plot displays the indicators in a circular format. Radial plots have radii that project from the centre such that the length of an individual radius corresponds to the measured value. Indicators all need to point the same way, so that high values of the indicator – and therefore long radii – always mean good performance. These values are sometimes connected to form an enclosed shape like a spider's web, hence the name spider plot (it's also known as a radar chart, a star chart and various other things). Figure 11.2 shows the results for an imaginary unit for the seven indicators in Figure 11.1.

The indicators in the spider plot can be ordered in different ways, and, depending on the consistency of the unit's performance, the unit can be made to look better or worse. The area within the web can be filled in if preferred. A second web can be added to show the results for another time period such as the previous year. Staffogia et al. (2011) give examples of and review variations on this theme: spider plots (ordered and unordered in terms of the performance on the set of indicators), target plots (again circular, but the scores are represented just as dots along each radius) and "spie" charts (which combine two superimposed pie charts). Target plots struggle to highlight small differences between indicators. The authors come out in favour of spie charts. In these, one pie chart acts as a base, and its partitions set the angle of each slice. The second pie chart is superimposed on the first, with the same angles for the partitions, but the indicator is expressed by changing the length of the radius. They also allow different weighting of the indicators, on the basis of predetermined criteria such as the number of patients affected or costs.

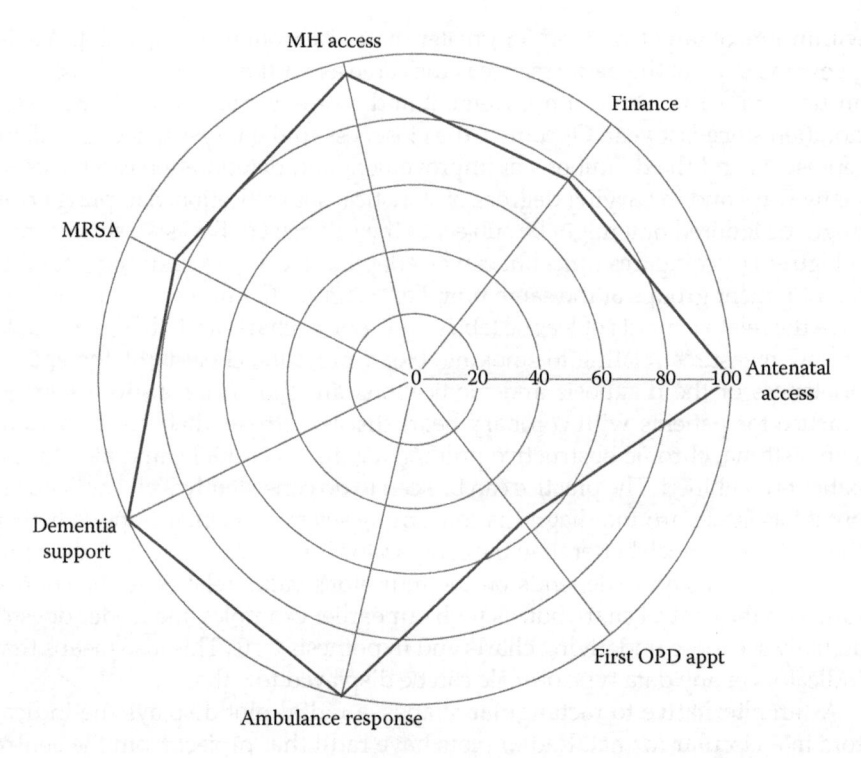

FIGURE 11.2
Spider plot of an imaginary unit's fictitious data.

11.2 Effect on Behaviour of the Choice of Format When Providing Performance Data

In this section, we consider the effect from two broad angles – staff and consumer – though there are going to be some similarities. It's easy for anyone to be overwhelmed by vast amounts of complex information even if we have some understanding of it. The research summarised in Daniel Kahneman's extraordinary 2011 book *Thinking, Fast and Slow* demonstrates the large number of temptations for our minds to take shortcuts and avoid difficult cogitation. Format matters a lot.

Of particular relevance here is the concept of value sensitivity: does our behaviour or our happiness change when the price or risk of something we care about changes? Examples of things we care about in this case include how much the care plan costs, how long we have to wait for the appointment, or what's the risk of a bad outcome following treatment. If we're willing to pay the same amount for a big diamond as a small one, then we are value insensitive regarding diamond size, which is the value here. A related concept is evaluability (Hsee and Zhang 2010). Suppose we learn that a medication that increases our risk of stroke by 2%. What do we do with this information? What does this actually mean for our health if we take the medication? Is our risk of stroke high enough now to worry about? Should we therefore consider not taking this medication? We might understand this fact as a concept in theory yet lack an emotional understanding of it. The single piece of information about the 2% higher risk of stroke lacks real meaning (evaluability) and so won't help our decision making (Hibbard and Peters 2003). For unit performance, we're likely to have more data to try to make sense of. If we try hard enough by slogging through existing information, we can create or determine meaning, but the temptation to avoid all that slogging is strong. Instead, we may prefer to rely on information that's more evaluable, that has an easier-to-grasp real meaning for us. In health contexts, money may be much more evaluable and easily understood than quality-of-care measures and will therefore win out in our decision making. This partly explains why many hospital visitors are so keen to know about car parking charges. Access, particularly measured in travel time or distance, is another such variable with good evaluability. This is perhaps one reason why patients tend to choose local hospitals over more distant ones even if the distant one achieves better outcomes – we find distance and travel times easier to grasp than performance measures. However, the good news is that the evaluability of information can be improved in a variety of ways, thereby making it easier for people to make more informed choices about the quality of services at different providers rather than just go for the nearest or cheapest. The evidence that this can happen generally comes from controlled studies in the psychology lab. An exception is a real-world study in south-central Wisconsin in which a hospital performance report using four different evaluable data display approaches was found to significantly change consumer views about the differences in quality between the hospitals. Those people seeing the report were more likely to recommend or choose one of the top tier hospitals than those not seeing the report (Hibbard et al. 2003).

Until they become patients themselves, healthcare staff don't have to choose which healthcare unit to attend for their medical problems. They do need to make sense of performance information either about themselves or their clinic, department, hospital or region. If the evaluability of this information is low, they may ignore it altogether in favour of something simpler to grasp. If ignoring it just isn't an option, as with an inspection report, they may come to the wrong decision. We'll now consider the method and impact of giving performance feedback to staff.

Feedback has been described as the cornerstone of effective clinical teaching (Hesketh and Laidlaw 2002). Without feedback, "good practice is not reinforced, poor performance is not corrected, and the path to improvement not identified" (Cantillon and Sargent 2008). Done badly, however, it can demotivate and even worsen performance. This is all true of students but also of anyone who is told how they are performing. Feedback should be part of an overarching quality improvement and physician education agenda rather than some isolated initiative scuppered by lack of resources: in short, effective feedback should be an integral part of clinical practice (Kaye et al. 2014).

Healthcare staff need different types and level of performance information depending on their role and responsibilities. Boards need information that is timely, reliable, comprehensive and suitable for board use. They have three core roles (Fresko and Rubenstein 2013). Their first core role is formulating strategy for the organisation, beginning with a clear plan to enable it to meet quality improvement challenges and a framework for a safe service, including safety ambitions such as zero harm and 100% reliability of care delivery (Berwick 2013). The board's second core role is ensuring accountability. The Berwick report, written in response to the scandal at Mid Staffordshire NHS Trust outlined in Chapter 2, set out the expectation that healthcare organisations "regularly review data and actions on quality, patient safety and continual improvement at their Board or leadership meetings". The third core role of the board is shaping a healthy culture for an organisation. This recognises that good governance comes from a shared ethos or culture and not just systems and processes, and a safety culture analysis and safety climate analysis will be needed (Vincent et al. 2013). In contrast, individual practitioners have quite different roles and responsibilities, which will be reflected in the type of performance information they will need and indeed receive.

A number of features of effective feedback have been identified. The review by Ivers et al. (2012) concluded that feedback may be more effective when performance is already low, when feedback is provided by a senior colleague or supervisor more than once verbally and in writing, and when there are explicit targets and action plans. Other factors found to be important in eliciting improvement following feedback are if information is provided close to the time of decision making and if the clinician had already agreed to review their practice (Mugford et al. 1991). Timely information is particularly important if there are safety risks that need urgent addressing (Vincent et al. 2013). Our group's work on giving anaesthetists benchmarked figures on post-operative measures such as pain and nausea identified via survey that, for clinicians to engage with effective quality monitoring and feedback, they need to believe that the indicators are relevant to them locally and they need to trust the credibility of the data (D'Lima et al. 2015).

Some barriers to effective feedback have also been found (Box 11.1). These include lack of measure prioritisation by the intended users; "measure fatigue" in addressing national, state and regional requirements (i.e., lack of alignment between the different levels of the system); the cost of measurement; the lack of timely data and the challenge of reliably measuring the performance of individual physicians and small physician groups given small denominator sizes for many measures (Damberg et al. 2012).

BOX 11.1 FEATURES OF AND BARRIERS TO EFFECTIVE FEEDBACK

Features:

- Given by senior colleague or supervisor
- Given more than once
- Verbal and in writing
- Explicit action plans and targets
- Timely
- Belief that information is relevant
- Trust in the data

Barriers:

- Information not given in an understandable manner
- Lack of measure prioritisation by the intended users
- Measure fatigue
- Data not timely or considered credible

11.3 The Importance of the Method of Presentation

There is some hard evidence that the method and format of the performance information matters. While the majority of this work has considered the public to be the main recipient, some studies have assessed the impact on staff.

11.3.1 Presenting Performance Data to Managers and Clinicians

In view of the unfortunate fact that league tables (see Section 9.1.3) used to be the staple of not just the media but also of the health service for presenting comparative performance results, Marshall et al. (2004b) assessed the decision making of UK directors of public health when presented with either league tables or control charts for each of three case studies. The performance data given to participants were either from real hospital data or, in the first case study, randomly generated so that there were in fact no outliers. League tables of death rates by unit were given in the form of a ranked histogram (or "caterpillar plot"), with the confidence interval superimposed on each surgeon's vertical bar. Control charts were of three types, but the accompanying text explained the control limits and the difference between "common cause" and "special cause" variation. The directors were randomly allocated to receive the same data in one of the two formats by mailed questionnaire, and 57 out of 122 responded. The main outcome measure was the percentage of participants who would take action on health service providers that they perceived to have unacceptable performance according to the information sent and who those providers would be. Those receiving the league tables labelled at least twice as many providers as requiring action than those receiving the control charts depending on the scenario. Respondents receiving data as league tables were significantly more likely to request further information on casemix. The authors concluded that using control charts rather than league tables for presenting comparative data would reduce over-investigations. This seems a reasonable conclusion, though the authors acknowledge that this was a simulation study, with the directors being asked what they would do in three hypothetical situations. However, the main limitation would seem to be the fact that the group given the control charts were told how to interpret them: "Data points within the control limits suggest common cause variation (intrinsic to the process). No action should be taken on these individuals: improvement can only be achieved by improving the process as a whole. Data outside the control limits suggest special cause variation and should be investigated." The accompanying text for the league table merely said: "The graph shows the percentage 30-day mortality rate. The bars are the 95% confidence interval for this result."

In a larger and broader study, Geraedts et al. (2012) sent a survey to 300 randomly recruited non-hospital-based physicians from all parts of Germany to determine which formats for presenting quality of care data are preferred by physicians who have to advise patients on which hospital to use. The clinical example chosen was knee replacements in seven fictitious hospitals. Six commonly reported quality indicators were provided: annual number of knee replacements; number of physicians in the department of orthopaedics; proportion of all physicians working in the department who were orthopaedic specialists; patient satisfaction (%); reoperations due to complications (%); and post-operative wound infections (%). For each format, participants had to name the overall best performing hospital for each of the formats and then the best and worst one on each indicator. The eight formats explored were as follows:

1. Ranked numeric table: Numeric data on all indicators. Hospitals are ranked by the sum of quality points achieved
2. Bar charts: Data on all indicators presented as bars (one bar chart per indicator)
3. Numeric table with traffic lights: Numeric data on all indicators supplemented by traffic light symbols
4. School grades table: Actual data replaced by school grades (one = very good to five = deficient)
5. Table with stars: Actual data replaced by one (worst) to five stars (best)

6. Comparative table: Actual data replaced by comparative terms (average, better, worse)
7. Traffic lights table: Actual data replaced by traffic light symbols (green = better, yellow = average, red = worse)
8. Simple star rating: Indicator data summarised to an overall star rating by dividing the sum of quality points achieved by six (indicators weighted equally)

Participants rated each format on a six-point ordinal scale from "insufficient" to "very good" and were asked two comprehension questions. Ranked numeric table and numeric table with traffic lights yielded "very good" or "good" ratings for all criteria. Comparative table came out worst. For the question "would you accept such a presentation format for choosing a hospital?" there were also significant differences: only numeric table with traffic lights achieved a clear majority of positive ratings (74%) and just 10% of physicians explicitly accepted the presentation formats comparative table and simple star rating.

The findings indicate that when choosing hospitals on the basis of comparative performance data, physicians prefer presentation formats that combine individual indicator values with evaluative cues such as rankings or traffic light symbols. They want the numbers but they also want some interpretative aids. Neither is sufficient by themselves, at least when they are advising patients on which hospital to choose for their surgery.

11.3.2 Presenting Results to the Public

The general tenet here seems to be "less is more". In 2002, the Centers for Medicare & Medicaid Services (CMS) launched the Nursing Home Quality Initiative, and one of its aims was to help people with Medicare and those who assist them make better decisions about their care. CMS rightly wanted to know whether the consumers understood the data as reported, but their commissioned studies and other works found that people had difficulty understanding bar charts and accessing the important information online. In view of this, Gerteis et al. (2007) assessed whether the adjunctive use of visual cues or tools other than bar charts would improve comprehension. When asked to choose a nursing home by comparing the CMS-reported quality measures, a sample of 90 consumers did indeed find tables with evaluative cues such as better, average or worse easier to understand than numeric tables and bar graphs. In particular, a format called numeric table with stars was highly preferred, as was evaluative table with words. Numeric table with stars shows the percentages with stars to indicate individual nursing home performance that is better than the state average. Evaluative table uses the words better, average and worse to indicate individual nursing home performance in comparison with the state average. In other works, people with low numeric skills, in particular, had been found to have the ability to pick out high-quality providers when information was structured according to its importance with the unnecessary stuff left out (Peters et al. 2007). The evidence from this and other works suggests that people are more motivated and better understand and use the information when cognitive effort is reduced and the meaning is made clear for them (Hibbard and Peters 2003).

Public reporting usually focuses on outcomes. Less attention has been given to the public reporting of costs. This is clearly relevant to some health systems more than others. What's useful to consumers depends on how much has already been taken care of via government spending or employment-based insurance schemes. Patients will be most concerned with out-of-pocket costs like co-payments or hospital parking fees. It's been argued that current ways of reporting costs are unlikely to drive improvements by patients choosing more efficient hospitals (Mehrotra et al. 2012). Many consumers believe that more care is better and that higher-cost providers are higher-quality providers: you get what you pay for. Mehrotra et al. make some recommendations for presenting cost data to the public. They give an illustrative scorecard for a set of maternity hospitals with just two columns of data next to the hospital name: quality of care (presented as a number of stars) and the out-of-pocket costs. Hospitals are sorted first by descending number of stars for quality and secondly by ascending cost.

The design principles for producing usable information hold true for any medium, but the web offers the most opportunities for interactivity and the use of multimedia effects. The web can also be used to tailor information or narratives to make the story more appropriate to the user's interests. Consumers with similar health conditions can be brought together for information sharing and emotional support, as reflected in the large number of condition-specific fora and online communities, often run by charities. Decision support tools could also help.

Important ethical considerations follow on from the findings of this research. Because the way health information is presented is very likely to influence how it's used, information producers have a responsibility to direct that influence in productive and defensible ways. Of course, the possibility remains for even the most conscientious producer to unintentionally manipulate people with negative results.

11.4 Examples of Giving Performance Information to Units

In addition to receiving performance information from accreditation bodies or regulators, units need to communicate such information to its staff to effect change. The balanced scorecard, popularised by Drs. Kaplan and Norton as a performance measurement framework (Kaplan and

Norton 1996), is a strategic planning and management system that is used extensively in business and industry, government and non-profit organisations worldwide. The term was coined in the 1990s, though the idea of a "dashboard" goes back earlier in the century to process engineering. Today, the balanced scorecard transforms an organisation's strategic plan from what can be a passive document into daily instructions for action. It therefore not only provides performance measurements but helps planners identify what should be done and measured. The "balance" is the provision of metrics beyond just the bottom line financial ones, as today's organisations recognise they also need to consider long-term capabilities and customer relationships. In healthcare, too, units need to be solvent but are judged on far more (often competing) criteria than that. The aim of the scorecard is to turn the team's vision and strategy for improvement into clear objectives and measures that can be monitored. Producing a balanced scorecard for an organisation as complex as a hospital is a major piece of work, and the task is usually devolved to directorate and department levels. The NHS Institute for Innovation and Improvement website (now the NHS Improving Quality) describes the steps involved in producing one, so that each of Kaplan and Norton's four areas – financial, customers, internal business processes, and learning and growth – have associated objectives, measures, targets and initiatives. For more information on the balanced scorecard, see the examples at the Balanced Scorecard Institute at http://balancedscorecard.org/Resources/Examples-Success-Stories such as how the Federal Ministry of Health in Ethiopia managed its organisational, political and infrastructure complexities to improve inpatient mortality and access to services. The pdfs also give examples of the scorecards themselves for different levels of the healthcare sector.

In many countries, the national regulator or accreditation body gives reports to units. National clinical audits are also common. In England, national and local clinical audit programs are overseen by the Healthcare Quality Improvement Partnership, an independent organisation led by the Academy of Medical Royal Colleges, The Royal College of Nursing and National Voices (a coalition of health and social care charities). The National Clinical Audit Programme comprises more than 30 clinical audits, covering a range of medical, surgical and mental health conditions. Some of these cover some, but not necessarily all, of the other countries within the United Kingdom. They all feed results back to units and publish annual reports, some of which contain unit-level performance results. The Danish National Indicator Project began by developing 6–10 indicators for each of eight diseases. Clinicians and managers in each unit and department receive monthly feedback of results. The hospital units can see whether they are below or above standards and whether their care is improving or worsening. Alongside this, a structured clinical audit process takes place nationally, regionally and locally to evaluate the results and plan improvements. Due to its success in raising standards across the country, the project has become a permanent part of the Danish healthcare system and has been expanded to other clinical areas.

Other than through national audits, performance information can be provided by professional societies or other non-governmental bodies. An example is the two Right Care Data Tools, the result of collaboration between the Royal College of Surgeons, Surgical Specialty Associations, NHS England and Right Care. These are intended to support the commissioning of elective surgery by local clinical commissioning groups (CCGs). Clicking on the Quality Dashboard at http://rcs.methods.co.uk/dashboards.html for NHS West Essex CCG brings up a list of surgical groups. Clicking on orthopaedics for knee gives several plots, of which the last two are shown in Figure 11.3. Measures are reported as numbers, on a horizontal "spine chart" that's essentially a one-unit funnel plot on its side, and as a trend. Accompanying documents give the metadata.

Another publicly accessible website that's aimed at commissioners is Public Health England's Fingertips (http://fingertips.phe.org.uk/).

Partial knee replacement (Partial knee replacement)

Metric	Period	Value	Mean	Chart	Trend
Age/Sex standardised activity (per 100,000 population)	RY Q1 1415	2.67	11.53		
Average length of stay (days)	RY Q1 1415	3.75	3.24		
7-day readmission rate (%)	RY Q1 1415	0.00	1.58		
30-day readmission rate (%)	RY Q1 1415	0.00	3.73		
30-day reoperation rate (%)	RY Q1 1415	0.00	0.69		
Day-case rate (%)	RY Q1 1415	0.00	1.25		

Total knee replacement (Total knee replacement)

Metric	Period	Value	Mean	Chart	Trend
Age/Sex standardised activity (per 100,000 population)	RY Q1 1415	132.15	128.08		
Average length of stay (days)	RY Q1 1415	4.57	4.72		
7-day readmission rate (%)	RY Q1 1415	9.52	6.35		
30-day readmission rate (%)	RY Q1 1415	24.29	15.09		
30-day reoperation rate (%)	RY Q1 1415	2.01	2.82		

FIGURE 11.3
Benchmarked performance for NHS West Essex CCG's commissioned elective knee surgery. (Taken from the Royal College of Surgeons' Right Care Data Tool, Data as on October 1, 2015. © Crown copyright. Reproduced with permission of Public Health England.)

Presenting the Results to Stakeholders

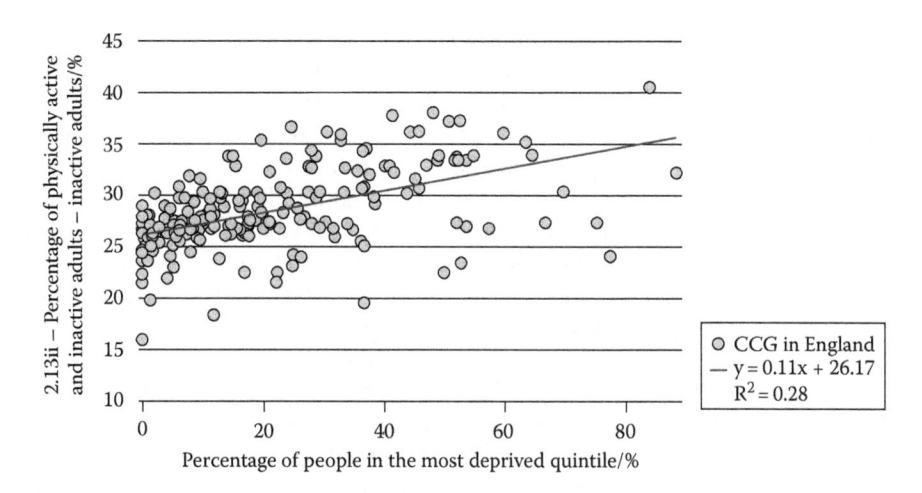

FIGURE 11.4
Example scatter plot from Public Health England's commissioning tool "Fingertips" showing CCG-level deprivation and adult inactivity. (© Crown copyright. Reproduced with permission of Public Health England.)

This covers a range of conditions and indicators. For example, the cardiovascular disease profile uses information from the National Cardiovascular Intelligence Network, which brings together a wide range of data on heart disease, stroke, diabetes and kidney disease for each clinical commissioning group (CCG) in England. Types of display include trend plot, funnel plot, spine chart, scatter plot and bar chart. The overall summary table gives the value for each indicator and CCG, with traffic light colour blocks to show the position relative to the benchmark. Users can choose between comparisons with the whole country and with the region. The scatter plot allows users to see the relation between two CCG-level indicators, for example, the percentage of people in the country's most socioeconomically deprived fifth and the proportion who are physically inactive (Figure 11.4). It even gives the R^2 and, if the R^2 is above 0.15, superimposes the trendline.

These two are not the only tools available for those that provide local services in England. A review by the King's Fund in 2015 (Ham et al. 2015) listed the main sources of CCG indicators and other indicator sets that have been used previously for commissioners as CCG Outcomes Indicator Set, Public Health England Fingertips, Commissioning for Value toolkit, HSCIC Indicator Portal, Better Care Better Value indicators, NHS Comparators and NHS Atlases of Variation. Behind these are the three NHS Outcome Frameworks and various policy documents for local and national priorities. There is also a lot of information on the performance of local areas rather than commissioners via Public Health England (on the health of local populations) and the national regulator, the Care Quality Commission (on the performance of providers). The King's Fund rightly concludes that this all needs simplifying and aligning.

When units receive data on their performance, what do they do? The collection of indicator data for audit, regulation or any other purpose incurs an administrative burden, but what's the most effective way to use those data for improvement? A systematic review of strategies for implementing quality indicators in hospital care and the effectiveness of these strategies covered 21 randomised controlled trials, controlled clinical trials and controlled before–after studies (de Vos et al. 2009). Most of these studies looked at process measures. The review found that the most frequently used implementation strategy was audit and feedback. This requires that a report including a summary of clinical performance over a specified period of time be given. Other strategies include educational meetings (e.g., training sessions, workshops), educational outreach (a trained independent person provided input), the development of a quality improvement plan and financial incentives for individual health professionals or institutions when they improve performance. In six out of nine studies that did not use feedback, the intervention was judged to be ineffective. Just three used feedback only, and two of these were judged ineffective. Of the nine studies that combined feedback with some other intervention, four were judged to be effective and another four partially effective. As we saw earlier in this chapter with staff needing information to advise their patients and with patients wanting to choose between providers, just giving the data is not enough.

In addition to government-run or otherwise "official" schemes, there has for some years been a thriving market for providing information to healthcare units. Companies such as 3M (based in the United States), Dr. Foster UK (now part of Telstra from Australia), de Praktijk Index (the Netherlands) and IASIST (Spain) are just four that offer benchmarking and various bespoke analytics for hospitals and/or other types of healthcare unit. Many of these companies offer sophisticated information tools whose details are proprietary and often not in the public domain. Information is generally presented in static formats, but to show changes over time, animations such as Gapminder can be very powerful. Its Gapminder World video plots the income per capita against life expectancy by country from 1800 to the present day and shows, for instance, that both have grown over time, beginning with income, and that inequalities in both have also become more apparent over time. In this era of Big Data, this market is only going to grow.

11.5 Examples of Giving Performance Information to the Public

In this section, we provide some screenshots from websites aimed primarily at the public. A selection of such sites from around the world is given in Table 11.2. They vary greatly in the quantity of data available, the type of

TABLE 11.2
Selected Public-Facing Websites for Obtaining Comparative Healthcare Unit Performance

Website Name	Country	URL
NHS Choices	England	http://www.nhs.uk/Pages/HomePage.aspx
NHS Inform	Scotland	http://www.nhsinform.co.uk/
Hospital Compare	United States	https://www.medicare.gov/hospitalcompare/search.html
Your Health System	Canada	http://yourhealthsystem.cihi.ca/hsp/indepth?lang=en#/
Australian Institute of Health and Welfare	Australia	http://www.aihw.gov.au/hospitals/
Ministerio de Sanidad, Servicios Sociales e Igualdad	Spain	http://www.msssi.gob.es/estadEstudios/estadisticas/sisInfSanSNS/home.htm

healthcare unit, the level of interactivity (some only give static reports or spreadsheets for viewing or downloading) and the amount of digging you have to do to get what you want.

Figure 11.5 shows the results from a search for general practitioners (GPs) in the small town of Epping in Essex, England, which yields the two practices in the town shown in the figure and others that are more distant and are not shown. The default sort is results by distance; other sorting options include user rating and whether the practice offers online appointment booking. The default view is key facts, as in Figure 11.5, but among the other views are patient experience scores, age of patients and use of hospitals.

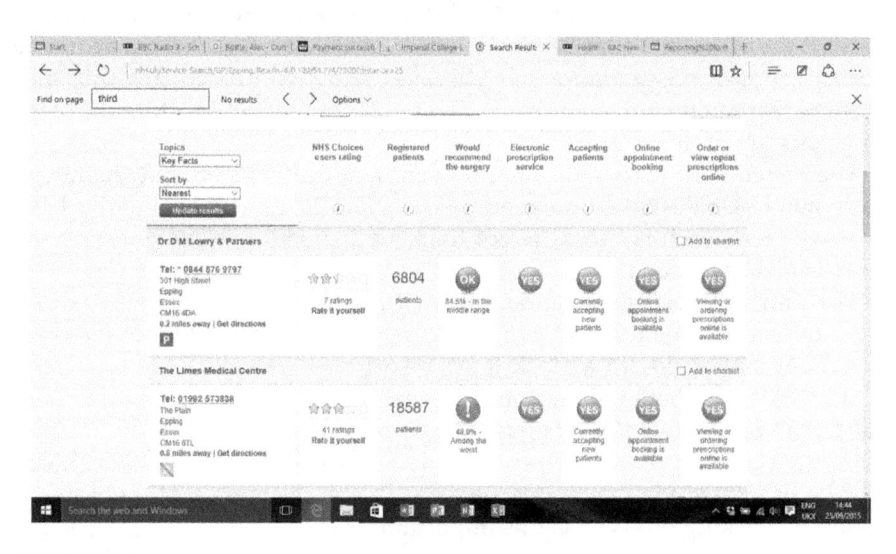

FIGURE 11.5
Screenshot from NHS Choices — Data as on September 25, 2015.

The NHS Choices users rating is how site users have rated the unit (in this case, practice). The proportion of patients who would recommend their GP surgery comes from the national biannual GP Patient Survey. The scores for this question are grouped into three bandings:

1. "Among the worst" if in the worst 25% of all scores nationally
2. "In the middle range" if in the middle 50% of scores nationally
3. "Among the best" if in the best 25% of scores nationally

This website provides numerical data and also puts labels alongside, using colour to reinforce the point. The Limes Medical Centre has a relatively low rating from the national Patient Survey (though it scores three stars out of five for its 41 NHS Choices users who scored it), which is marked by a white exclamation mark on a red background as if as a warning.

In the United States, the main source of official hospital performance data for Medicare patients is Hospital Compare. Typing "Seattle" into the first box on Hospital Compare's search page yielded 18 results, ordered by distance from the centre of the city. The initial results page gives just the basic information about the hospital such as its address, type and whether it has emergency services. You can then click on up to three to compare. Clicking on "Compare now" brought up a series of measures, beginning with "general information" that includes a couple of rather specific data points such as whether a safe surgery checklist is used. Measure categories that each have their own tab are general information, survey of patients' experiences, timely and effective care, complications, readmissions and deaths, use of medical imaging, and payment and value of care. For example, under readmissions and deaths, figures are available for the chosen hospitals for five diseases and two procedures (hip and knee replacements are combined). Figure 11.6 shows the results for readmissions and deaths for COPD.

The benchmark (national) rate for each measure is given, but the hospitals' own risk-adjusted rates are not in this view. Instead, an evaluative label is used. Here, each is "no different than the national rate". Requesting the graphs gives Figure 11.7.

The national average is marked on with a vertical dotted line, and the rates for each hospital are shown by a caret mark with confidence intervals extending either side in a block that's a sort of dirty yellow colour. Green would indicate "better than national rate" and red "worse than national rate". The number of patients used in the calculation is also given. Further information about the measures, such as the denominator and data source (the latter located via the "about the data" link at the bottom of the page), is found by clicking on "why is this important?"

Presenting the Results to Stakeholders 205

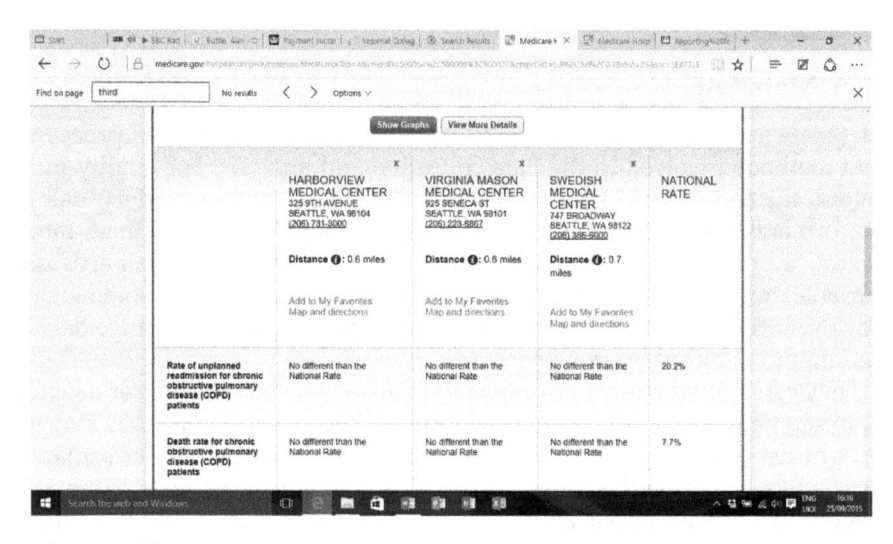

FIGURE 11.6
Illustrative Hospital Compare results for three Seattle hospitals for readmissions and deaths in patients with COPD.

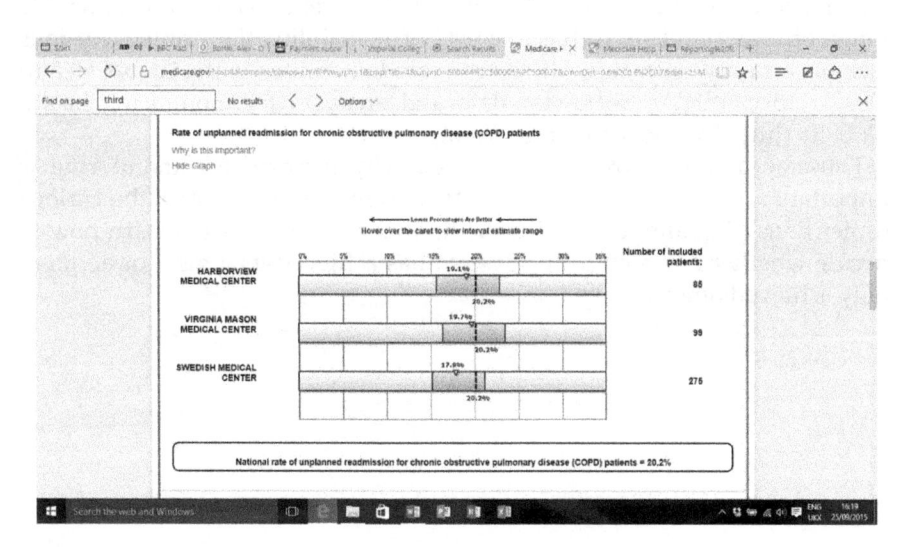

FIGURE 11.7
COPD readmission rates for three Seattle hospitals displayed as bar charts by Hospital Compare.

11.6 Metadata

Metadata are "data about data" or information about the purpose, processes and methods involved in the data collection and analysis. For quality indicators, the term "technical specifications" is also common. Every use of performance information needs some level of metadata to help them interpret it, even if only methodologists dig into the details of how particular variables are coded and processed. Such detail is necessary somewhere, however, to ensure transparency and repeatability and to encourage debate and methodological improvement.

They can appear alongside graphs or tables as labels, footnotes or legends or in separate sections or documents; online they can also appear as mouse-over boxes or pop-up windows. The level of detail put by the performance data will depend on the intended users' expected level of knowledge and interest, the relative importance of the data and the metadata, and how much room there is. If the data can more or less stand alone with little explanatory text beyond a table or graph title and axis labels, then the technical details can be moved elsewhere. This is the case for the examples presented in this chapter. A common statistical decision is whether to include confidence intervals or other uncertainty estimates around lines on a graph or columns of figures in a table. Statistical rigour sometimes needs to give way to readability. For example, in the WHO's World Health Statistics reports, the data for some countries are quite flimsy, but the reports aim to cover a wide range of topics and so, in the interests of readability, the printed versions lack uncertainty estimates. These are provided online via the Global Health Observatory, WHO's portal for data and analyses for monitoring health globally (http://www.who.int/gho/en/).

Thus far in this book, we've covered why performance monitoring is important and how to measure, monitor and report indicators of the various dimensions of quality of care. In the final main chapter, we'll learn how to decide whether our monitoring system has achieved its aims – or accidentally achieved others.

12

Evaluating the Monitoring System

Aims of This Chapter
- To list the main elements of an evaluation of a performance monitoring system
- To outline the statistical methods for such evaluation

Monitoring healthcare performance is in general complicated and expensive, so we need to know whether it achieves what we hope it achieves without too much in the way of adverse effects. As we saw in earlier chapters, performance targets and incentive schemes distort behaviour, from the submission of suspicious-looking data (e.g., peak in emergency department waiting times just before the four-hour target) to taking our eyes off the ball regarding non-incentivised areas (e.g., those conditions not covered by P4P schemes). It's vital to capture such unintended consequences in the evaluation. Regarding the intended consequences, part of the evaluation will consist of a before versus after comparison of indicators using the same quantitative data as in the monitoring. As such data are unlikely to tell the whole story, the other part will consist of the collection and analysis of qualitative and/or additional survey data on factors such as staff attitudes and patient experience. Introducing a monitoring system will lead to far-reaching effects on the healthcare system and as such is akin to introducing a complex healthcare intervention. Examples of such interventions include redesigning a service involving multiple providers or efforts to improve public health by a national screening or health promotion campaign.

While randomised controlled trials (RCTs) are sometimes possible and have been done, most evaluations use an observational study design such as a controlled before versus after comparison. This chapter outlines the study design and statistical methods. We'll also briefly outline the elements of an economic evaluation, though this is not intended to be an economics textbook. We conclude with some remarks on the importance of qualitative methods to obtain a full picture of whether the monitoring scheme has been a success and, as with economics, suggest some further reading.

Some of the examples that we use in this chapter to illustrate the methods are not from the evaluation or monitoring literature. This is because, as far as we're aware, not all the methods that we present have been used to assess monitoring schemes yet. We nonetheless include them as they have a lot to offer such assessments.

12.1 Study Design and Statistical Approaches to Evaluating a Monitoring System

If targets or standards are set, then we can test whether they're met. With baseline data before the introduction of the monitoring scheme, we can compare the degree to which the targets or standards are met before versus after its introduction. In the absence of pre-existing trends, this is straightforward and can be as simple as a chi-squared test to compare two proportions. This "uncontrolled before and after comparison" design has been the typical approach for the studies assessing the impact of accreditation. A 2011 Cochrane review of randomised controlled trials (RCTs), controlled clinical trials, interrupted time-series and controlled before and after studies evaluating the effect of external inspection found only two such studies (Flodgren et al. 2011): the South African RCT described in Section 2.4.2 and a controlled before and after comparison. Despite its RCT design, evidence from the former study was downgraded in terms of quality using the GRADE scoring system (see http://clinicalevidence.bmj.com/x/set/static/ebm/learn/665072.html) due to concerns over imprecision and publication bias.

The statistical methods follow on from the study design. Traditionally, despite the problems with the South African study just mentioned, the RCT has been regarded as the gold standard design. It's the most effective way of removing confounding. There are two broad types, which represent the ends of a continuum: pragmatic and explanatory (terms coined in the 1960s). Explanatory trials aim to test whether an intervention works under optimal situations, and so assess the intervention's *efficacy*. Inclusion criteria are stringent and conditions are controlled to study and maximise the effect, but the results don't always translate well to the real world of clinical practice. In contrast, pragmatic trials aim to test whether an intervention works in real-life routine practice conditions, and so assess the intervention's *effectiveness*. With non-randomised designs, treatment selection bias can occur due to physician and patient preferences. It can be more extreme within some patient subgroups, for example, the elderly, and it can't be overcome by increasing the sample size. Clinical characteristics, such as disease severity and patient frailty, relate to both treatment selection and outcome but might not be recorded in sufficient detail even by clinical databases to allow traditional methods of adjustment (regression or propensity-score analysis: see the following text).

In many quality improvement (QI) projects, however – and one can easily argue that performance monitoring fits the description of one – other paradigms are recognised as having some key advantages over the RCT despite the greater risk of residual confounding. Population-level interventions can be difficult to manipulate experimentally, will be hard to justify ethically if it has known benefits (as the control group, if one can be found, would be denied them) and can be politically problematic.

We should note in passing that there has been a good deal of methodological research into RCT designs. For instance, an adaptive trial technique known as response-adaptive randomisation allows the random allocation ratios to change over time according to how well each treatment being tested is performing. If the new treatment is doing better, then fewer new patients will be assigned to the other one that's faring worse. A JAMA Viewpoint piece suggests how the best elements of RCTs and observational studies might in future be combined so that trials are also quality improvement projects (Angus 2015).

Natural experiments occur in a variety of settings and are useful alternatives when there are such doubts over the effects of the intervention or when RCTs are impractical or unethical. When a policy is introduced nationally, then the only option is to use another country as the control, which is clearly problematic. The United Kingdom, however, is four countries in one, and there have been several studies of the impact of a policy change that occurred only in one of the four, using the others as controls. This is an example of a natural experiment. The healthcare systems in the four countries are not identical but are very similar – each has a National Health Service (NHS). In 1998, the Labour government devolved power to an elected parliament in Scotland and elected assemblies in each of Wales and Northern Ireland. Each of the four countries took different paths regarding their NHS policy. A 2005 study used published indicators on health and healthcare spending, staff and activity to compare each country in 1996/7 before devolution with 2002/3 after devolution (Alvarez-Rosete et al. 2005). Some of the information is collected in similar ways in each country, though there are some differences regarding, for example, operation waiting time definitions. Data were presented without confidence intervals or significance tests, presumably because they were national and therefore assumed to be known without sampling error. This is rarely done but makes for easy interpretation of tables. The authors noted that the most striking differences between the countries since 1998 related to waiting times for the first outpatient appointment and elective operations. The large falls in England were due to the setting and implementing of quantitative, time-specific targets, with strong performance management and sanctions for failure.

12.1.1 Interrupted Time-Series Design and Analysis (*)

The second study included in the Cochrane Review of accreditation evaluation mentioned earlier concerned the Healthcare Commissions Inspection Programme (OPM Evaluation Team 2009), which included all acute hospital

Trusts in England. The standards in this program were developed and launched by the Department of Health, which in 2006 enacted the new legislation with the aim to reduce the number of healthcare-acquired infections. The evaluation of the program considered a range of measures, such as hospital Trust compliance with the code of practice and patient and public confidence in healthcare. Here, though, we'll focus on the rates of hospital-acquired methicillin-resistant *Staphylococcus aureus* (MRSA) infections to illustrate a key approach that has a number of advantages over the simple uncontrolled before versus after comparisons: interrupted time series (ITS) design. The MRSA rate is mandatory for Trusts to report each quarter, and data were available for one year before the initiation of the inspection process and for two years after. This allowed re-analysis by the Cochrane reviewers, who found statistically non-significant effects on MRSA rates of the Inspection Programme. After that review, Devkaran and O'Farrell (2014) presented an ITS analysis of the impact of hospital accreditation on clinical documentation compliance in the UAE. To illustrate the method, we will look at this study in some detail.

Their data were in the form of an interrupted time series, meaning that they were divided into segments of time, which gives rise to the alternative name of segmented regression analysis. With one intervention, as is typically the case, the segments correspond to pre- and post-intervention time periods. Each segment of time has an associated mean level and a slope to be estimated. There are three main components in the output from an ITS model: pre-intervention slope, step change in level immediately following the intervention and change in the slope post-intervention. We're generally interested in changes in means (i.e., in slopes and levels) but the method can also be used to look for changes in variance, which can be of interest. A readable account of this technique is given by Wagner et al. (2002). Figure 12.1 shows the segments.

A key advantage of the ITS approach over simple before versus after analyses is its accounting for pre-existing time trends so that potential critics can't argue that things were getting better anyway before the intervention. As one should do with every time series analysis, Devkaran and O'Farrell were careful to inspect their data for features such as trend, seasonality, outliers, turning points and/or sudden discontinuities. Time-series data generally have at least one systematic pattern, of which the most common are trends and seasonality. Trends are generally linear or quadratic and are often found by using plotting moving averages or by regression. Seasonality is a trend that repeats itself systematically over time. To check for linear trends, the authors applied ordinary least-squares regression analysis, using the Durbin–Watson statistic to test for the presence of autocorrelation. Autocorrelation – when the data at one time point correlate at least partly with the data at neighbouring time points – is common in time-series data and violates the assumption of independence of error. If ignored, this will inflate the chance of seeing an association that doesn't

Evaluating the Monitoring System

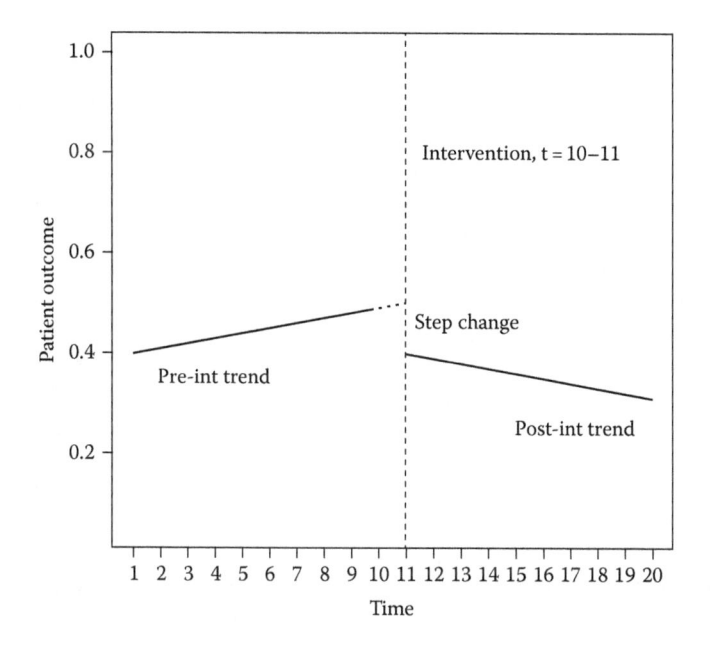

FIGURE 12.1
Time segments pre- and post-intervention in an interrupted time series design.

actually exist (the type I error rate). The authors used the Dickey–Fuller statistic to test for seasonality and stationarity: a common assumption in time-series methods, this holds that the mean, variance and autocorrelation structure don't change over time. These problems are typically dealt with by taking the difference of the series from one period to the next and then analysing this differenced series. This differencing might need to be done more than once to remove the problem. For simple seasonal patterns, dummy variables may be included, for example, winter versus nonwinter. Finally, once all the statistical assumptions have been investigated and checked, the model's output can be interpreted. While ITS has been described as the most powerful, quasi-experimental design to evaluate longitudinal effects of time-limited interventions (Wagner et al. 2002), the interventions need to be considered in the context of whatever else is happening around the same time. Healthcare policy changes rarely happen in isolation. The authors considered whether the hospital had seen changes in management or structure during the study and also noted that there was no monitoring by Joint Commission International at that time.

With ITS, the intervention is assumed to occur completely at one time point and its effect noted straightaway by the step change in level, by a change in slope, or both. There are clearly two big assumptions in this. With complex interventions like many quality improvement projects or the introduction of a performance monitoring system, it's often optimistic to suppose that rollout occurred everywhere within a short space of time. Policies tend to have early adopters and laggards. If the intervention simply takes several time points to be rolled out, then this is straightforward to handle a priori in the data and modelling. If the intervention takes time to have an effect, then sensitivity analyses can be used to artificially "extend" the rollout time or to introduce a third segment, giving "pre-intervention", "post-intervention pre-effect" and "effect" segments.

ITS practitioners are often asked by those thinking of using the method how many data points are needed to estimate the trends. You always need more data points than parameters, but beyond that, it's hard to answer beyond the irritating "it depends". One rule of thumb states that at least 25 observations (time points) before and an equal number after the intervention are needed in order to produce reliable results (Johnston et al. 1995). The more parameters and the greater the randomness (noise) in the data, the more points you need. If you've nothing to adjust for beyond time and the start of the intervention, and if in the simple plot of outcomes over time the data look so linear that a child could accurately draw a line through the points, then you could get away with just three points for each segment. To make simple adjustments for month or quarter using dummy variables, then the bare minimum is 6 observations for quarterly data and 14 observations for monthly data. Other time-series methods for dealing with trends need a few more parameters and hence observations. However, if the trend isn't linear and the fit isn't good, then you might not have controlled for the trend sufficiently. In this situation, some of the apparent effect of the intervention could be due to the uncontrolled pre-existing trend. With greater randomness, the question of how many observations becomes hard to answer simply.

12.1.2 Adjusting for Confounding (*)

This vital task can be divided into adjusting for *observable* confounding and adjusting for *unobservable* confounding. The first of these is easier, simply because, as the name implies, we have information on the confounders. There are three approaches to adjusting for these:

1. Matching
2. Regression
3. Propensity scores

Matching is common in case–control studies, particularly by age and sex. For each patient in one group, we seek one or more patients with the same matching characteristics. This leads to the same distribution of, for example, age among cases and controls, thereby controlling for age; in case–control studies, this has the extra benefit of a gain in efficiency. Regression was discussed in detail in Chapter 6 on casemix adjustment and is the most common approach to the problem. It works well if two assumptions are met: (1) we've measured all the important factors and (2) we've correctly specified the relations between factor and outcome. If the second of these is not met, then we've got all the important factors but they're distributed markedly differently in the intervention and control groups. One use of propensity scores is to address this for multiple potential confounders via propensity score matching (PSM). A propensity score is the estimated probability of being "exposed" to an intervention given a set of covariates. For example, suppose we're interested in whether patients get better outcomes with laparoscopic or open colorectal surgery and need to account for the difference in the distributions of patient age in the two groups. In a non-randomised study, patient selection effects will explain some of the difference in outcomes, for instance, if younger patients are more often chosen for the laparoscopic approach. It's commonly estimated for each patient by logistic regression with the group (intervention or control) as the Y variable, which in this example would be laparoscopic surgery. Factors such as age, comorbidity and disease severity would go into the model as covariates. Propensity scores are useful if the distributions of the factors vary by group or if we are uncertain about the form (linear, S-shaped, etc.) of the true relation between the factor and the outcome. Given the rise in popularity of PSM, it's worth describing the process and briefly reviewing the evidence for its effectiveness.

For PSM, the procedure is as follows:

1. Develop a logistic regression model for predicting the intervention as a function of the risk factors (don't include factors that are not associated with the outcome): the output probability is the propensity score.
2. Rank the set of patients from highest to lowest propensity score.
3. Divide into subsets with similar scores: for example, quintiles work well.
4. Within each subset (quintile), divide patients into intervention and control groups based on which group they actually belonged to.
5. Sample equal numbers of intervention and control patients from each quintile (this is known as pair matching).
6. Examine the distribution of risk factors in this sample between the two groups.
7. Fit new model: raw and adjusted estimated intervention effects should be similar due to the matching, but this will get rid of residual differences.

The pair matching can be done in various ways, such as optimal matching and greedy nearest neighbour matching (particularly within certain "calliper" widths); there are also versions of many-to-one matching that we won't consider here. Optimal matching forms matched pairs so as to minimise the average within-pair difference in propensity scores. In contrast, greedy nearest-neighbour matching first selects a treated subject; the untreated subject whose propensity score is closest to that of the treated subject is chosen as its matched control. In our example, "treated" would refer to the laparoscopic group and "untreated" to the open surgery group. Using Monte Carlo simulation, Austin (2014) compared 12 algorithms including variations on these two approaches such as with the use of "callipers" (a certain amount of latitude regarding how near the control's propensity score had to be to that of the treated subject). These specified calliper widths impose a maximum difference in propensity scores between treated and untreated subjects within a matched pair. Without this condition, it's possible for some matched pairs to have quite different propensity scores. The use of the callipers therefore tended to result in less biased estimates compared with the other matching algorithms. On the other hand, the use of callipers can lead to the exclusion of some treated subjects from the matched sample if there's no untreated subject with a propensity score that's near enough. This will have implications for the generalisability of the results. Optimal matching and nearest-neighbour matching both result in all treated subjects being included in the matched sample and thereby avoid this potential bias.

There are four methods in which the propensity score can be used, and the various ways of matching on it as just described make up just one. The others are stratification on it, covariate adjustment using it (the simplest and most common way in the medical literature), and inverse probability of treatment weighting (IPTW) using it. With IPTW, observations are downweighted rather than discarded. Austin (2009) compared the four methods and found that they all reduced a big proportion of the systematic differences in patient characteristics between the treatment groups. In simulation and a case study, PSM and IPTW beat stratification and covariate adjustment, eliminating virtually all observed systematic differences.

More difficult but still vital is to try to consider potential confounders for which we lack data entirely: *unobservable* confounders. PSM can't help us with this. They may be out of sight, but they should not be out of mind. Bias will also remain if our observed confounders are imperfectly measured. Under certain circumstances, that is, with certain strong assumptions, other approaches can be tried in order to deal with the unobserved factors, in particular difference-in-differences (DID) analysis, instrumental variable (IV) analysis, and regression discontinuity (RD) designs. We will now consider these three in brief.

12.1.3 Difference-in-Differences

Rather than compare a measure over time between exposed and unexposed groups, this method compares the *change* over time between them. The differencing procedure controls for unobserved individual differences and for common trends. It assumes that the unobserved characteristics are fixed and that the outcomes in each group would change in the same way in the absence of the intervention. This means it's vulnerable to changes in the composition of the groups and to any external influences that affect the exposed group in a different way from the unexposed group. Nonetheless, it's widely used, particularly in economics, and should be considered.

Suppose that some states or regions adopt some legislation, which could of course be a monitoring system, in a given year (the "intervention"), whereas others don't. We have data for the before and for the after period for each state. In terms of notation, the simplest case is when we have the same number of observations on the same people before and after, but neither of these conditions is at all necessary for us to use this technique. If we're happy to assume that, in the absence of the intervention, the unobserved differences between intervention and control groups are the same over time, then we could use data on intervention and control group before the intervention to estimate the "normal" difference between intervention and control group. Then we compare this "normal" difference with the difference after the receipt of the intervention. In other words, the difference of the before and after period for the adopting states minus the difference of the before and after period for the non-adopting states yields the DID estimator.

An example is the comparison of outcomes in stroke patients following the 2010 centralisation of services in two regions of England, Greater Manchester and London, as had occurred in a number of other countries (Morris et al. 2014). Concerns about the number of patients being transported greater distances and other issues meant that the Greater Manchester model differed from the hub-and-spoke model in London, so these two regions were not combined. Two outcomes were considered: mortality and length of stay. Each was adjusted for age, sex, comorbidity and several other factors using regression. Of note is their reasonable decision to use only patients who had a stroke before the reorganisations so that the risk adjustment was "not contaminated by the [service] changes". The resulting regression coefficients were used to predict the probability of mortality for every patient in periods before and after implementation. These were aggregated to create a data set of the actual percentage of patients who died and the expected percentage who died by admitting hospital and quarter. Expected mean length of stay was estimated using a generalised linear model with the gamma family and log link to account for the typical skewness in this outcome measure. Hospitals and quarters were included as dummy variables (fixed effects), though by using linear time trends and interaction terms they also checked whether risk-adjusted mortality and length of hospital stay had a different linear trend in Greater Manchester and London compared with the rest of England before the service reconfigurations (these were all non-significant). The DID analysis showed that the change in mortality in Greater Manchester was similar to that in the rest of England, but in London it was significantly greater. For example, within 30 days of admission for stroke, patients in London had a –1.3% greater change in mortality (95% CI –2.2 to –0.4, $p = 0.005$) compared with the change in the rest of England, that is, a 1.3% bigger fall. The authors converted these absolute differences into relative reductions in mortality and into numbers of deaths: 1.3% change in the absolute risk represents a relative reduction of 7% and 194 fewer deaths (95% CI 60–328). In both areas, there was a significantly larger decline in risk-adjusted length of hospital stay compared with the rest of England. They again converted the DID estimates into relative changes and total number of outcomes, in this case bed days. In the Discussion section, they considered alternative explanations for their findings. The hospital administrative database that was used lacked information on stroke severity. The national audit for stroke showed that severity does indeed differ between Greater Manchester, London and the rest of the country, but what is more important is whether this changes sufficiently over time to be able to account for the apparent improvements in patient outcomes: though there were no clear trends, the authors did not rule out stroke severity as a confounder. The database also only captured in-hospital activity, so would have missed anyone dying of stroke without being admitted, for instance, if ambulance travel times had lengthened appreciably: figures from the ambulance service suggest that this was not the case.

12.1.4 Instrumental Variable Analysis

An instrumental variable is a factor, such as treatment assignment in a randomised controlled trial, which is associated with the outcomes of interest only via its association with exposure to the intervention. It also has to be independent of other factors associated with that exposure. Its occurrence creates a natural experiment, doing the job of randomisation. An example should make things clearer.

In observational studies of hospital treatment, the distance between the patient's home and the hospital is often used as an instrument. For this to be valid, it needs to correlate with the probability of getting that treatment but be uncorrelated with health status (i.e., other predictors of the outcome of interest). For example, McClellan et al. (1994) set out to determine the effect of more intensive treatments on mortality in elderly patients with acute myocardial infarction. This was an observational study using administrative Medicare claims data. The treatment of interest was cardiac catheterisation, a marker for aggressive care, and the instrument was defined by asking whether the nearest hospital to the patient's home was a catheterisation hospital: differential distance = distance to nearest catheterisation minus distance to nearest non-catheterisation hospital, based on the zip codes of residence and hospital. To be a good instrument, it needed to pass these two tests:

1. Is it correlated with treatment (cardiac catheterisation)?
2. Is it uncorrelated with patient severity (other predictors of the outcome besides treatment) that is observable in claims data?

There were 26.2% who received catheterisation if the nearest was a catheterisation hospital, and 19.5% if not, so it passed the first test of being correlated with the treatment. The differential distance was found to be unrelated to age and comorbid disease, so the instrument seemed to pass the second test too. As the authors put it, "Thus, differential distances approximately randomise patients to different likelihoods of receiving intensive treatments". They found that catheterisation was associated with a 5%–10% point reduction in mortality, nearly all on the first day.

More recently, in a comparison of two kinds of stents in patients aged ≥85 years undergoing percutaneous coronary intervention, Yeh et al. (2014) used quarterly drug-eluting stent (DES) use rates as an instrumental variable (IV). Unadjusted one year mortality rates were 14.5% for DES versus 23.0% for bare metal stents, a large difference that was greatly at odds with RCT results and considered implausible. In contrast, using IV analysis, DES was associated with no difference in one year mortality ($p = 0.76$) or bleeding ($p = 0.33$) and with significant reduction in target vessel revascularisation (risk difference −8.3%, $p < 0.0001$). See their paper for the technical details, including how they tested those underlying assumptions that could be tested.

As with any method, the IV approach has its proponents and its uses. It's dependent on the existence of suitable instruments and the validity of the accompanying assumptions, some of which unfortunately can't be tested.

12.1.5 Regression Discontinuity Designs

This approach, which is still undergoing methodological development nearly 60 years after its invention, exploits a step change or "cutoff" in a continuous variable used to assign treatment or determine exposure to an intervention. The assumption is that units (patients, individuals, areas, etc.) just below and just above this threshold will otherwise be similar in terms of characteristics that may influence the outcomes of interest. If this is true, an estimate of the treatment effect can be obtained by comparing regression coefficients on either side of the cutoff. Developed specifically to address the special case of subject assignment based on a pretest score, it's been popular in economics since the late 1990s, and technical readers of that persuasion should see the guide by Imbens and Lemieux (2008). It's fair to say that it's been used considerably less in medicine and epidemiology. Most of the 32 articles in a recent systematic review of RD in that space via PubMed analysed the effectiveness of social policies or mental health interventions (Moscoe et al. 2015).

Whenever a decision rule assigns treatment, such as antihypertensive therapies, to patients who score higher (or lower) than a particular cutoff value on a continuously measured variable, such as blood pressure, RD can be used to estimate the causal effect of the treatment on outcomes. Such cutoffs or thresholds for treatment occur a lot in medicine. Treatment assignment following such a rule can be either deterministic (every patient on the one side of the cutoff value receives the treatment and every patient on the other side does not) or probabilistic (the probability of receiving the treatment is higher on one side of the cutoff value than on the other side). The first case is called sharp RD and the second fuzzy RD.

Under certain conditions, it's possible to infer that a difference in outcomes is the causal result of the assignment variable's cutoff point. Thinking has changed since the early days of the method from considering global mean treatment effects to local mean treatment effects at the threshold. As we approach the cutoff value from above and below, the patients in the two groups become more and more alike on both observable and unobservable characteristics; in a small area around the threshold, the only difference is in treatment assignment. This is often true in clinical settings, where simple measurement error, such as with blood pressure or concentration of some physiological quantity such as cholesterol, means that whether a person near the cutoff falls above or below it is essentially random. If that's the case, then we have essentially replicated randomisation as in an RCT. Three conditions need to be met for RD:

1. The decision rule and cutoff value to be used to assign treatment status are known.
2. The assignment variable is continuous near the cutoff value.
3. Potential outcomes (or their rates if using binary outcomes) are continuous at the threshold.

The decision rule needs to be strictly implemented. If it is not, for instance if people's preferences or other considerations mean that some patients are misclassified, then selection bias is introduced and the treatment effect is likely to be biased; the same is true in RCTs if the randomisation is incomplete or compromised. Note that measurement error in the assignment variable does not cause bias. The reason for this is that, once the observed assignment variable scores are known, treatment assignment is completely determined and hence independent of anything else, including the perfectly measured true assignment variable scores. Similarly, in RCTs, treatment assignment is completely determined by the randomisation scheme and hence independent of anything else. Regression to the mean, which is due to measurement error in the covariate, is also not a problem here in that it does not affect the treatment effect (it does bias the estimate of the covariate effect, though, but that's usually of less interest).

The fuzzy RD design is similar to a randomised experiment with imperfect treatment compliance. The average causal treatment effect that fuzzy RD gives us is equivalent to the intention-to-treat effect from trials. Analysis of fuzzy RDs is similar to that for sharp RDs, but with an extra step, the inclusion of an instrumental variable based on treatment assignment. This instrument lacks the usual weakness of instruments, which is the reliance on the untestable assumption that the instrument is as good as randomly assigned.

It's useful to begin the RD analysis by plotting the assignment variable against the probability of treatment status, which displays the distribution of treatment assignment. In the example of blood pressure and antihypertensive medication, we'd plot the blood pressure against the probability of being given the medication. If the probabilities of being assigned to a given treatment are all either 0 or 1, we have sharp RD. If not, we have the fuzzy version. For most evaluations of large-scale public health programs, the fuzzy form of RD will be used because the cutoff will only determine treatment status for some patients. We won't go into further details of the method here. For step-by-step instructions for implementing RD, see the review and guide by Moscoe et al. (2015).

However, as with everything there are cons as well as pros to RD. It's essentially just a before and after test and can't address complex time patterns. This means that RD applications will generally need to be followed up using other approaches to determine how long the effects of an intervention last. Its assumptions are weaker than those of other methods, but they still need to be met. Zuckerman et al. (2006) list these briefly as follows: "an exclusive cutoff criterion should be employed; the pre-post distribution can be described as a polynomial function rather than logarithmic, exponential, or some other function; a sufficient number of pretest values should be present; all subjects must come from a single continuous pretest distribution; and the intervention program was uniformly delivered to all recipients."

In contrast to RD, interrupted time series has been widely used to estimate the effects of policy changes and has more advocates. As seen in the earlier section, ITS is a similar design to RD. In the most rigorous ITS scenario when a policy is implemented very rapidly and short-term outcomes are assessed at frequent intervals, ITS can be interpreted as a subtype of the RD design (Moscoe et al. 2015). Calendar time is the assignment variable and the cutoff occurs when a new policy is implemented. This is presumably why, for example, the method for the evaluation of hospital report cards for trauma by Glance et al. (2014) was described as RD by the authors and ITS in the accompanying editorial (Dimick and Hendren 2014). They're both right.

The 2012 Medical Research Council guidelines for using natural experiments to evaluate population health interventions (Craig et al. 2012) conclude that, because of the strong, untestable assumptions and data limitations, these methods for dealing with unobserved confounders – difference-in-differences, instrumental variables and regression discontinuity – "are therefore best used in conjunction with additional tests for the plausibility of any causal inferences". What this means in practice is that studies should be replicated with subsequent careful synthesis (where possible) and that the results from different approaches should be compared. The guidelines paper does, however, conclude that, "More could be learnt from natural experiments in future as experience of promising but lesser used methods accumulates". I agree with this. Although ITS is well established, the time for RD does not yet seem to have come in health services research. As has been pointed out, there was 30-year lag in widespread acceptance of the randomised experiment in social and health sciences research (Shadish et al. 2002). We shouldn't be surprised that non-randomised designs still encounter resistance.

12.2 Economic Evaluation Methods

As noted in Chapter 2, regulating and monitoring healthcare services and systems are expensive. Just as many quality improvement initiatives have a built-in financial assessment to give a more complete picture of their worth, so one ought to do the same for a monitoring scheme, which can be regarded as a quality improvement (QI) intervention. If the results of the monitoring are followed by positive action by healthcare providers to better their performance, then we have something to put in the "benefits" box. If no such positive action results, then we'll only have an entry for the "debits" box. A QI initiative that leads to no improvement is still a QI initiative, just not a very successful one.

The aim of this section is just to give a flavour of the different types of perspective and analytical approaches on offer. There have been fairly few economic evaluations of performing monitoring systems, an exception being the approach taken for an evaluation of an accreditation program in Australia. We list the steps in performing the most common type, a cost-effectiveness analysis (CEA), but for detailed information on health economics we give some guides and textbooks at the end.

There are several different types of economic evaluation available. It can be defined as "the comparative analysis of alternative courses of action in terms of both their costs and their consequences". We therefore need to identify, measure, value and then compare. With a national non-phased rollout of the monitoring scheme, the only available comparator will be historical, that is, the state of the country's health just before rollout. If there's no real comparison element, then we have a description rather than an evaluation. Box 12.1 outlines the different options, adapted from Drummond et al. (2005).

BOX 12.1   BASIC TYPES OF ECONOMIC ANALYSIS

1. No comparison group
   Cost description: Looks at costs only, not consequences
   Outcome description: Looks at consequences only, not costs
   Cost–outcome description: Looks at both

2. Comparison group
   Cost analysis: Looks at costs only, not consequences
   Effectiveness evaluation: Looks at consequences only, not costs
   Cost-effectiveness analysis: Compares cost per unit of consequence, for example, life-year gained
   Cost–utility analysis: Compares cost per unit of quality of life, for example, QALY gained
   Cost–benefit analysis: Compares costs and consequences, both measured as money

The choice of method will be influenced by data availability and the desired definition of value, but it might also be possible to employ more than one in the same project.

The most common type of economic evaluation for evaluating health interventions is CEA. This compares the costs and outcomes of two or more alternative interventions, such as the monitoring scheme versus the status quo. These costs (net costs) could be the cost per disease avoided, cost per death avoided or cost per additional expected life year. The net cost includes the cost of implementing the monitoring scheme minus the costs that are avoided because of the beneficial effects of the scheme. A ratio is calculated for each alternative intervention: the numerator is the cost in dollars, and the denominator is the measurable health outcome. The intervention with the lowest dollar value per health benefit is the more cost-effective of the set of alternatives.

The steps involved in a CEA, many of which are common to the other types, can be outlined as follows. First, set up the comparison:

1. Define the problem and objectives – economic evaluations can be done for a number of different reasons, as we will discuss shortly.
2. Identify the comparator, which for a monitoring system will be the status quo, either historically or in comparable regions or healthcare units that are not part of the system.
3. Describe each group for comparison in detail, which could just be the monitoring system and the status quo. Identify all the inputs required to implement each group.
4. Define the perspective of the CEA. The choice of perspective influences which costs to consider and whose value to use. There are three basic viewpoints that are used in economic evaluations: a societal perspective, the provider's perspective or the private (patient or household) perspective (see the following text).

Next, identify, collect and bring together the data:

5. Identify, measure and put values on costs.
6. Identify and measure effectiveness: measures for this are likely to be the same as those used to measure the impact of the monitoring system, so when you design the impact assessment analysis plan, also consider the CEA element.
7. Discount future costs and effectiveness for studies whose costs and effects occur over several years.
8. Run sensitivity analysis: this means re-estimating the model using minimum and maximum ranges for some indicators where the true values are unknown or where there's likely to be a lot of variability in the estimate. It's especially important to conduct sensitivity

analysis for those variables that are expected to have a large impact on the cost-effectiveness of the intervention.
9. Finally, use the CEA results in decision-making. This is the point of the data collection and analysis.

Whereas a cost-effectiveness analysis estimates the benefit of an intervention simply in terms of units of effect, such as years of life gained or hospital bed days saved, a cost-utility analysis (CUA) estimates the benefit using the broader measure, utility. This puts a value on each unit of effect, so a year of life gained is adjusted for the value we put on that year and turned into a quality-adjusted life-year (QALY). If someone gains a year of life thanks to the intervention but is physically or mentally impaired enough to reduce the value of that life from their perspective, then they haven't gained a full year. The big advantage in this approach is that different types of effect can be compared in non-monetary terms. To do this, we need a common unit of currency, which is often the QALY; we met the concept of value in health and the measures DALY and QALY in Chapter 4. The approach then becomes a special case of CEA, where the health outcome to be compared among the alternative interventions is the QALY.

The third major option, a cost–benefit analysis (CBA), converts the health gain into money, which means that not only can it compare two or more interventions, it gives us information on the absolute benefit of each one, that is, the resources used by the intervention compared with the resources saved (or created) by that intervention. Brent (2003) argues that cost-effectiveness analysis, cost-minimisation and cost-utility analysis are essentially subtypes or "shortcuts" within CBA. This is because the CUA is a special case of the CEA that uses QALYs as the outcome measure, and CBA takes CEA one step further by converting the outcomes in CEA into monetary terms.

In addition to different basic types of economic analysis, there are different types of perspective. Economic evaluations are often used to assess the relative efficiency of alternative healthcare interventions, so they commonly use the health service's perspective. This is sometimes called the decision-maker's perspective. The decision-maker focuses only on the resources under their direct control in a fixed timeframe, so the only relevant costs are those met from the fixed health budget. However, maximising health outcomes within the limited healthcare budget would not necessarily maximise the welfare of society within all of its resources available (gross national product). One reason is that non-healthcare sectors could incur costs or benefits as a result of healthcare interventions. Another reason is that alternative uses for resources outside the healthcare sector may yield greater benefit to society: this is what the concept of opportunity cost captures. As health economics is concerned with society's welfare and not just one of its systems, a societal perspective is recommended by the WHO and others (Tan-Torres Edejer et al. 2003). On page 19 of their guide, they define this to mean "that all costs should be included regardless of who pays them, and resources used or created by health interventions should be valued at the benefit foregone because society could not use the resources in their next best use. This next-best use might be in health or non-health".

Thus far, there has been little work done to economically evaluate monitoring schemes such as accreditation. Mumford et al. (2013) drew up a protocol to "determine and make explicit the costs and benefits of Australian acute care accreditation". They proposed a cost–benefit analysis (CBA). Unlike with cost-utility analysis and cost-utility analysis as noted earlier, with CBA both the costs and the benefits are expressed in monetary terms. The Australian National Safety and Quality Health Service standards are a mix of organisational and clinical outcomes, which is common in many countries, so the authors argue that converting them all into monetary terms is the most appropriate option. It compares all benefits against all costs. The resulting cost–benefit ratio suggests whether or not the benefits outweigh the costs of an intervention, and hence CBA provides a broad societal perspective. Modelling the costs and benefits of a complex intervention such as accreditation is a huge task, and the paper outlines their planned approach. As with any economic evaluation, a number of sensitivity analyses are proposed. Putting a dollar value on health benefits, which is a necessary step in CBA, is fraught with conceptual and empirical difficulties. This helps to explain the popularity of other methods.

Drummond et al. (2005) emphasise that the types of analysis in Box 12.1 all generate quantitative data about the costs and consequences of a given healthcare intervention, but they're best thought of as "methods of critical thinking" to help the decision-maker's judgment rather than remove their responsibility. Their book covers the analytical methods and types of economic data in readable detail and includes worked examples from published studies as tutorials. In the words of Robert Brent, editor of a book on CBA that uses the same structure, it "opens up the field to non-economists" (Brent 2003). We direct the interested reader to those two books. For CEA, see the WHO's guide by Tan-Torres Edejer et al. (2003).

This concludes our coverage of statistical methods relevant to healthcare performance monitoring. For the full picture on whether the monitoring scheme has succeeded, however, qualitative methods are needed as well as statistical and economics viewpoints. These tackle the how and why of decision making. The value of qualitative methods in outcomes research and healthcare evaluation is being increasingly recognised. Assessing aspects such as organisational change, teamwork and clinical leadership is hard to do well with numbers. The most common research method is the interview but also frequently used are group discussions and observation of participants. The approach is often exploratory, starting with observations and developing hypotheses (inductive reasoning) rather than testing hypotheses using observations (deductive reasoning). Combining quantitative and qualitative data is known as a mixed methods evaluation; how exactly these two very different elements should be done is the subject of ongoing research. For complex healthcare interventions such as monitoring systems, detailed knowledge of not just the intervention but also the context are vital in order to interpret the evidence of the intervention's effectiveness. If the numbers are disappointing, assuming that we've picked the most appropriate study design and comparison groups, we need to know whether it's because the intervention is flawed in its concept or theory, whether it was badly delivered (poor implementation) or even whether it was the right thing done well but just in the wrong place (wrong context). Translating and scaling up healthcare quality improvement and patient safety interventions are still hard, largely because of context. A qualitative method taken from the social sciences into health, ethnography, is a promising way to dig in and try to understand the professional, organisational and cultural aspects of context. For more information on how to do it and the challenges involved, see the narrative review by Leslie et al. (2014). For a summary of what qualitative methods in general have to offer outcomes research, see, for instance, Curry et al. (2009).

13

Concluding Thoughts

Statistics has often been called an art and a science. Likewise, performance monitoring requires the combination of methodological rigour with judgment to navigate our way through a series of competing priorities and trade-offs. Table 13.1 sets out what we consider to be the main dichotomies that we've discussed in this book. They can be distilled into two conflicts between simple and complex and between specific and general. The best answer often lies somewhere in-between.

13.1 Simple versus Complex

At one extreme, then, we have an indicator devised by a single person in an afternoon, calculated on the latest data already on that person's computer, and put out the same week. No complicated, costly expert panels or stakeholder opinion-testing. No waiting for more detailed data collection. No doing lots of sensitivity analyses or anything lengthier than basic quality assurance checking. This information is very timely and yet low cost. On the other hand, it might be wrong with serious consequences for some organisations incorrectly labelled as outliers or it might simply be of little interest to the intended users.

At the other extreme, we have an indicator developed following a full review of available evidence, preliminary definition and analyses, extensive stakeholder engagement and consultation, revised definitions, the de novo collection of sophisticated data, sufficient analysis to explore and minimise the risk of biases due to factors such as data errors and unmeasured confounding, pilot testing of the final version, polished quality assurance and, finally, release. This information has the full support of all stakeholders, comes from reliable data and is scientifically unimpeachable. On the other hand, it's cost a fortune and is out of date.

TABLE 13.1

Trade-Offs in Healthcare Performance Monitoring

Point	Counterpoint	Comments
Scientific rigour	Understandable and usable	Do you want the best model that fancy methods can deliver that the users don't trust because they don't understand it – or a simpler but inferior one that users understand (or claim to) and are therefore more likely to use?
Cheap but dirty	Clean but expensive	Re data limitations: you can have something now for cheap that's noisy and not quite what you want – or you can pay and wait for something better.
Thorough	Timely	Relates to the previous pairs. Rapid analysis can incur errors and misinterpretation, but feedback must be timely to be effective.
Whole patient pathway	Healthcare silos	Most indicators relate only to a specific part of the healthcare system: easy but narrow.
Whole-patient	Disease-specific	Many indicators, the practical reality of treatment and the underpinning evidence base mostly assume that patients have just one disease, but many don't.
People with diseases	People with life needs	Patients have diseases, or at least symptoms, but their goals are expressed in terms of what they can do, e.g., go out with friends and play with the grandchildren, rather than biomedical parameters like FEV1.
Important	Actionable	Outcome versus process measures: outcomes are important, but we improve care through improving processes.
Specific patients	General population	Individual versus population: tailored info on limited subset of patients and/or their care/illness versus less specific measures for large groups of patients and/or care/illnesses.

13.2 Specific versus General

With this pair, we have at one extreme an indicator that fully acknowledges that each person is an individual and that their medical, social, psychological and other needs are unique. The indicator combines their diabetes, angina and occasional migraines, their limited social network, shyness and difficulty getting long-term employment, and their desire for advice on sustainable weight loss, better time management and help with applying for child support. This information captures all the patients' main worries and shows exactly where the healthcare provider is going wrong or doing well, but is relevant to about six people in the whole country.

The converse would be an indicator that covers everyone in the country, such as life expectancy from birth. Everyone's interested in that, but what do you do with this information to try to improve the figures? There's also a lot more to life than being alive for three score years and ten.

13.3 The Future

One of our subject's challenges is to ameliorate these tensions between the simple and the complex, the specific and the general. The rise of Big Data (and more importantly, access to it) and personalised medicine will give us ever greater understanding of disease, its treatment and its outcomes and ever more sophisticated information on all three. Indicators of the future could include not just if a patient with a given set of diseases has had all the processes met for all their diseases' clinical guidelines, but also if they've had their treatment tailored sufficiently to their phenotype and genotype – and if they've had good outcomes, not just within one healthcare silo such as the intraoperative period but for their whole patient journey, and not just in a biomedical, physical sense but also in a social and psychological sense.

We expect risk-stratification and decision support tools for bedside clinical management and patient counselling to improve considerably. Performance information could become much more real time, with dynamic risk-adjustment algorithms taking account of the latest temporal changes in benchmark risk and sitting on top of clinic and hospital IT systems instead of static models applied to data from months or even years earlier. Priorities for areas for quality improvement that lead to the greatest overall good – or some other desirable criterion that society decides – could be identified from an overarching analytical framework that could be tailored to account for government and regulator directives and patient values alike. Performance monitoring is hard but it's here to stay across the world and will only get bigger.

Appendix A: Glossary of Main Statistical Terms Used

Most of the statistical terms used in this book relate to risk-adjustment models or to statistical process control (SPC), with a few that are neither. We have, therefore, organised the terms in this appendix into three categories. This appendix is aimed at readers who are not very familiar with statistical methods and terms.

A.1 Terms Concerning Risk-Adjustment Models

There are two principal, related risk-adjustment modelling approaches – *standardisation*, which can be direct or indirect, and *regression*. The steps for the calculation of SMRs using standardisation were given in Chapter 7. Briefly, with the indirect method, the expected numbers of deaths are derived by multiplying the unit's population by the national rate. In the direct method, it's the other way round: the unit's rates are multiplied by the national population to get the expected deaths, which are then divided by the actual number of deaths nationally. Regression is much more flexible in terms of the types of outcome and casemix variables that can be accommodated.

Regression at its simplest is an equation that describes a straight line via two coefficients: the slope and the intercept with the y-axis. With something such as blood pressure that's normally distributed – or can be trimmed, transformed and otherwise persuaded to look fairly normal – we only need this straight line, where the slope of the line describes, for example, how BP changes with age. We have the following linear model:

Predicted BP = Estimated BP at age zero (the intercept)
 + Slope for age in years

To get the predicted BP for someone, take their age in years and multiply by the value of the slope. Then, add the value of the intercept. This all assumes that the BP goes up neatly in a straight line with age. This might not be the case. Age is often not treated as a continuous variable but instead divided into groups such as five-year bands. Suppose we have three age groups: under-45, 54–64 and 65+. The equation is the same in principle but has different coefficients. In order to solve the equation, we need to pick one age group as the "reference" group, for example, the under-45s. This means that its effect (mean BP in the under-45s) is represented by the intercept. The algorithm can then generate a coefficient for each of the other two age groups:

Predicted BP = Estimated BP due to being in age 0–44 (the intercept)
+ Extra BP due to being in age 45–64 *or* Extra BP due to being in age 65+

Everyone's predicted BP is the sum of the intercept term and one of the other two terms.

For a binary outcome, we can't just use Y as 0 or 1 as the maths wouldn't work. To make it work, we need a way of transforming the Y variable so that we can draw a straight line with its intercept and slope as given earlier. The usual approach is to use the natural logarithm of the odds, or the "log odds" for short: take $\log(P/(1-P))$, where P is the probability (risk) of death. The *logistic regression* model is

Predicted log odds of death = Estimated log odds due to being in age 0–44 (the intercept) + Extra log odds due to being in age 45–64 *or* Extra log odds due to being in age 65+

A person's predicted log odds of death is the sum of the intercept term and one of the other terms. Adjustment of risk for other factors such as sex and comorbidity is done by adding more terms to this equation. To convert the log odds into a probability, a simple formula is applied.

Logistic regression is more similar to indirect standardisation than it might appear at first glance. Instead of simply multiplying to get the expected counts for each age group, each patient gets an estimate probability of death, which can then just be summed across the age groups to get the total expected for the unit. These probabilities are estimated by a procedure called maximum likelihood, which is a matter of calculus to find the most likely values for the coefficients for age given the data.

In these examples, age is handled as a *fixed effect*. Each age gets its own coefficient which in these examples where age is the only variable in the model, is the same for everyone within the same age group. Another common way to think about fixed effects is that they are interesting in themselves rather than representing a larger population of members (in this case, age groups). In contrast, *random effects* are considered to represent a larger population with its own distribution of effects (often taken as a normal distribution with mean effect of zero). They can be included in *multilevel models* and are therefore associated with *shrinkage*, all of which we now explain.

Multilevel or "hierarchical" models are commonly used to deal with the *clustering* of patients within units, as is the case when, for example, a set of patients are all treated by the same physician or attend the same hospital (see later text). Here, we can either treat a small set of physicians as

Appendix A: Glossary of Main Statistical Terms Used 231

being of particular interest, for example, we can compare Dr. Wang and Dr. Patel directly as fixed effects, or we can treat the whole set of physicians in our data set as random effects, representing a larger population of physicians who might also see the same kinds of patients in the same healthcare setting.

In the *fixed effects* case, Dr. Wang and Dr. Patel get their own coefficient as per age group given earlier. If they are the only two physicians being compared, then the outcomes for one, for example, Dr. Wang, will be incorporated into the intercept, and we'll get a coefficient for the other physician, for example, Dr. Patel, which represents the effect of being treated by them over and above the effect of being treated by the other. Note that the coefficient can be negative so that Dr. Patel could have worse outcomes than Dr. Wang. If they have no deaths in their patients, then their estimated SMR will be zero. In the *random effects* case, however, this is no longer so. Their zero observed deaths will be replaced by an estimated – "shrunken" – value that considers all physicians in the data set. The fewer patients they see, the more their SMR will be pulled towards the overall mean, that is, 100.

The choice of whether to implement a multilevel model is complicated. Some people take into account the amount of *clustering*, which is how similar the outcomes are in patients who are treated by the same physician. Strong physician effects will result in more clustering. This can be seen by exceptionally good or bad surgeons in operations and outcomes that are highly sensitive to how good the surgeon is. A measure of clustering is the intraclass correlation coefficient (ICC). The higher this is, the more you should consider taking it into account. In linear models, where the Y variable is normally distributed, its calculation is more straightforward than in non-linear models such as logistic regression, where the outcome measure isn't normally distributed until it's transformed (by taking the log-odds for a binary variable such as death). There are ways round this problem, though, that we won't go into here.

How do we know that our risk-adjustment model performs well? There are various common measures. For linear models, the R^2 is common. This is the proportion of the variation in the outcome that the model (the set of risk factors) explains. For non-linear models, the maths is again trickier, though versions of this measure exist. For binary outcomes and logistic regression, two other measures are more common. One is the area under the receiver operating characteristic curve, otherwise known as the *c statistic*, or *discrimination*. This ranges from 0.5 to up to 1 (1 is not always the theoretical maximum, despite what you might read). The c statistic tells us how good the model is at distinguishing between people who die and people who survive. It doesn't tell us whether the model guesses very accurately – that's the *calibration*, as commonly summarised by the Hosmer–Lemeshow chi-squared statistic and test. A well-calibrated model would predict exactly the right number of deaths in all sets of patients, for example, in both low-risk and high-risk patients. What this means is that, for any subgroup of actual risk, the number of expected deaths should equal the number of observed deaths.

The bigger the difference between the observed and expected deaths within each risk subgroup, the worse the calibration.

A more sophisticated approach than standardisation or regression is to use a *machine learning* method. There is now a vast array of these in gainful employment in medical applications. The oldest is the *artificial neural network*, with its layers of artificial neurons and nodes that connect them: an input layer for the predictor values, a set of nodes that convert the inputs into the output and an output layer that gives us its results. There are different ways of setting these up, but the most common way in risk prediction or risk adjustment is to get the computer to make an initial guess, compare it with reality (e.g., whether the set of patients actually lived or died), feed back to the network the results of how close the first guess got, incorporate these results and try again. In this way, the network "learns" what's the best set of outputs (calculus is involved) for the data at hand. A common measure of performance is the mean squared error. Many of the other methods have analogous concepts. Machine learning can get extremely technical, but there are a growing number of books and videos aimed at people without a background in linear algebra or optimisation.

A.2 Terms Concerning Statistical Process Control

Statistical process control (SPC) is a family of methods, from the humble *run chart* through to multivariate *control charts*. The run chart is just a plot of data points over time, for example, average waiting times by week. Various rules such as those set out by the Western Electric Company determine whether there are any "unusual" patterns in the data, that is, whether there is "special cause variation" or "common cause variation". The latter is just random variation or sampling error (noise) that we want to filter out in order to spot the former. Examples on a run chart of special cause variation include trends and long runs (e.g., eight consecutive points on the same side as the overall median). Run charts are simple but, like most simple things, are limited.

Performance monitoring is more likely to use control charts. These employ *control limits* to spot unusual data points, which run charts can't do. Control limits are often placed at three standard deviations either side of the mean; 95% control limits are common too and map to roughly two standard deviations. Any point lying above the upper or below the lower control limit is labelled an *outlier* (it shows special cause variation). By chance, if all units showed only random variation (if they were all "in control"), then 95% of such units would lie within 95% control limits, so 5% would be labelled as outliers. 99.8% control limits are commonly used to reduce the chance of false positives. With these limits, just 0.2% of in-control units would be labelled as outliers. *Funnel plots* are commonly used to display the performance of multiple units on the same chart and often employ 95% "warning" limits and 99.8% "alarm" limits.

Sometimes, we draw a funnel plot or some other way of showing the variation between units and see rather more than 5% of units outside the 95% control limits, for instance. This is called "overdispersion", which means more variation than one would expect just due to chance. Causes included dodgy data, a poorly thought-through indicator, insufficient risk adjustment and real variations in quality of care. Sadly, it's somewhat challenging to tell the difference and usually impossible from the chart alone.

Some SPC charts incorporate not just the current patient or observation but also information from previous patients or observations. Examples include the cumulative sum (CUSUM), which weights the current and all previous patients equally, and the EWMA, which downweights more ancient ones. The chart statistic is some function of the difference between the observed and expected and is plotted over time. If it crosses some preset *threshold*, then the chart is said to *signal* (or to alert) and the signal (or "alarm") represents special cause variation. All alarms require investigation, as we can't tell just by looking at the chart what the reason for the alarm is. If the threshold is set too low, then you'll get lots of false alarms. If set too high, then detection of real problems will be impaired. The performance of such charts is often described by the false alarm rate (FAR) and the successful detection rate (SDR), though many other measures have been suggested.

The FAR is the proportion of units whose performance is in fact average that signal. For charts that incorporate prior information into the chart statistic that's plotted, such as CUSUMs and EWMAs, this needs to be given for a given length of monitoring such as per 1000 patients. This is because the longer the monitoring, the greater the chance of an alarm. Eventually, given long enough, all charts will signal and the FAR will then be 100%.

The SDR is the proportion of units whose performance is not consistent with the average – they're outliers, either good or bad – who signal. Again, this is expressed for a given length of monitoring for CUSUMs and EWMAs. A low threshold will raise the SDR but also raise the FAR. A high threshold will reduce both.

A.3 Other Terms

A.3.1 Imputation

With missing data, we can either drop those patients entirely, drop those variables from the model (e.g., if it's a risk-adjustment model) or plug the gaps (impute values). Imputation is done by averaging information from patients who don't have missing data. If we don't think the missing value depends on other variables (the so-called missing completely at random), then single imputation can be used. A common example of this is *last observation carried forward*, when we simply take the last known value for, for example, the patient's BMI at their last weight check and assume that, at the time point when we don't know their BMI, it hasn't changed since the last time we did know it. Often, however, missing data are at best *missing at random*, when they do depend on other variables. In this instance, we can fill in the gaps by what we know from those other variables using *multiple imputation*, a procedure that involves regression. This has been shown to work well.

A.3.2 Multiple Testing

Like medical tests, every statistical test has an inherent false alarm rate. When we pick $p < 0.05$ as the conventional cutoff value in our definition of statistical significance, we are accepting a 5% chance of a false alarm. In a statistical hypothesis test, we set up a null hypothesis and look to our data for evidence that the null hypothesis is false. The null hypothesis typically involves no difference, for example, Clinic A's vaccination rate is no different from the average vaccination rate. Note that the p value is *not* the probability that the null is true. Instead, it measures how compatible your data are with the null. It's the probability of obtaining an effect at least as extreme as the one in your sample data, assuming that the null is true. Let's say that Clinic A's rate was 80% and the average rate was 90%, giving a p value for this comparison of 0.03. Does this suggest that its rate is low or could it just be due to random variation when sampling patients? If we go with a 5% cutoff, we'd conclude that its rate differs from the average. This 0.03 is the chance of seeing a rate of 80% or lower when in fact Clinic A's rate was 90%. 0.03 is not that likely, but it's not zero. It could yet turn out to have the average rate, which would mean that this is a false positive or false alarm. If we take another sample of patients, maybe things will look different.

If we look at not just its vaccination rate, but a whole load of other measures, we could find that some give us significant differences and others don't. Without further investigation, we can't tell which are false alarms and which are real differences, though really low p values suggest the latter. The response to reducing the chances of false alarms is lower the cutoff, making the test more stringent. This problem of multiple testing is extreme in healthcare performance monitoring because we have not just multiple performance measures but also multiple units – and multiple time points. Everyone involved needs to understand that false alarms are inevitable.

There is a philosophical divide in statistics between the "Bayesians" and the "Frequentists" (non-Bayesians) that some people like to make a lot of but most tend to shrug off – and many are in both "camps" depending on the real-world problem at hand that needs answering. Thomas Bayes, a nineteenth-century English vicar, dabbled in probability theory and produced a formula that describes the chances of two interrelated things happening. It's used a lot in the quantitative sciences and can also help with interpreting evidence in lawsuits.

Suppose we need to estimate the risk of something happening, and we have some data. Frequentists define probability as the long-run frequency of a certain measurement or observation. With a small sample, we obtain an imprecise estimate. We can see how imprecise it is by constructing a 95% confidence interval (CI) around our estimate. For a small sample, this will be wide to reflect our uncertainty. What we need are more data to get closer to the truth, that is, to the true value of the risk. This will give us a narrower 95% CI. In contrast, in the Bayesian framework, there is no true value out there to be found. Instead, there are data which we can use as evidence for particular hypotheses, such as the risk of interest is less than 10%. We also need to think about and then incorporate our prior beliefs about this risk of interest. If we want to know whether it will rain today, we could use past data to say what proportion of days in this country has it rained. It doesn't matter that it's rained every day for the past fortnight: what matters is only the overall risk of rain in the past. To be Bayesian, we could add in the recent persistent dampness as well as the past overall risk of rain into our model. The two estimates can differ considerably. Of course, both could be right and both could be wrong.

In Chapter 7, we discussed whether a surgeon who performs only a few operations is likely to have an average complication rate or a higher-than-average complication rate. The frequentist would simply use available data to estimate the rate and compare it with the national average to see if it's high or not. The Bayesian could bring their knowledge of the commonly observed relation that performance improves with increasing number of operations that a surgeon does. The surgeon is thereby assumed to have a higher complication rate. Their data may then suggest otherwise, but if they don't have many data they will be stuck with the higher rate – guilty until proven otherwise. We can still calculate a measure of our uncertainty around the surgeon's rate, called a 95% credible interval in this framework. Strictly speaking, it's not the same as a confidence interval, but we still all prefer narrow ones to wide ones.

The two approaches can have different methods, and there is specialised software to implement Bayesian analysis, but really the two differ in their goals. One is not better than the other. Hammers are not inherently superior to screwdrivers, and it's useful to have both in your toolbox.

References

Adams JL. (2009) *The Reliability of Provider Profiling. A Tutorial.* Santa Monica, CA: RAND Corporation. Available at http://www.rand.org/content/dam/rand/pubs/technical_reports/2009/RAND_TR653.pdf (accessed September 2015).

Agency for Healthcare Research and Quality. (2008) Patient safety indicators (PSI) composite measure workgroup final report March 2008. Available at http://www.qualityindicators.ahrq.gov/Downloads/Modules/PSI/PSI_Composite_Development.pdf (accessed September 2015).

Alexandrescu R, Bottle A, Jarman B, Aylin P. (2014) Classifying hospitals as mortality outliers: Logistic versus hierarchical logistic models. *Journal of Medical Systems* 38(5):29.

Alkhenizan A, Shaw C. (2011) Impact of accreditation on the quality of healthcare services: A systematic review of the literature. *Annals of Saudi Medicine* 31(4):407–416.

Almoudaris AM, Burns EM, Bottle A, Aylin P, Darzi A, Faiz O. (2011) A colorectal perspective on voluntary submission of outcome data to clinical registries. *British Journal of Surgery* 98(1):132–139.

Alt FB. (1985) Multivariate quality control. In: Johnson NL and Kotz S. (eds), *Encyclopedia of Statistical Sciences*, Vol. 6. New York: Wiley.

Alvarez-Rosete A, Bevan G, Mays N, Dixon J. (2005) Diverging policy across the UK NHS: What is the impact? *British Medical Journal* 331:946–950.

Andersen B. (1990) *Methodological Errors in Medical Research.* Oxford, UK: Blackwell.

Andrianopoulos N, Jolley D, Evans SM, Brand CA, Cameron PA. (2012) Application of variable life adjusted displays (VLAD) on Victorian admitted episodes dataset (VAED). *BMC Health Services Research* 12:278.

Angus D. (2015) Fusing randomized trials with big data. The key to self-learning health care systems? *Journal of the American Medical Association* 314(8):767–768.

Armstrong RA. (2013) Statistical guidelines for the analysis of data obtained from one or both eyes. *Ophthalmic and Physiological Optics* 33:7–14.

Arsenault J, Michel P, Berke O, Ravel A, Gosselin P. (2013) How to choose geographical units in ecological studies: Proposal and application to campylobacteriosis. *Spatial and Spatiotemporal Epidemiology* 7:11–24.

Ash A, Shwartz M. (1999) R2: A useful measure of model performance when predicting a dichotomous outcome. *Statistics in Medicine* 18(4):375–384.

Ash AS, Fienberg SE, Louis TA, Normand S-LT, Stukel TA, Utts J. (2012) Statistical issues in assessing hospital performance. Report for the Committee of Presidents of Statistical Societies. Centers for Medicare & Medicaid Services. Available at https://www.cms.gov/Medicare/Quality-Initiatives-Patient-Assessment-Instruments/HospitalQualityInits/Downloads/Statistical-Issues-in-Assessing-Hospital-Performance.pdf (accessed September 2015).

Ash AS, Shwartz M, Pekoz EA, Hanchate AD. (2013) Comparing outcomes across providers. In: Iezzoni LI (ed.), *Risk Adjustment for Measuring Health Care Outcomes*, 4th edn. Chicago, IL: Health Administration Press, p. 362.

Ashcroft DM, Morecroft C, Parker D, Noyce P. (2005) Safety culture assessment in community pharmacy: Development, face validity and feasibility of the Manchester Patient Safety Assessment Framework. *Quality and Safety in Health Care* 14(6):417–421.

Asplund K, Bonita R, Kuulasmaa K, Rajakangas AM, Schaedlich H, Suzuki K, Thorvaldsen P, Tuomilehto J. (1995) Multinational comparisons of stroke epidemiology. Evaluation of case ascertainment in the WHO MONICA Stroke Study. World Health Organization Monitoring Trends and Determinants in Cardiovascular Disease. *Stroke* 26(3):355–360.

Austin JM, D'Andrea G, Birkmeyer JD, Leape LL, Milstein A, Pronovost PJ, Romano PS, Singer SJ, Vogus TJ, Wachter RM. (2014) Safety in numbers: The development of Leapfrog's composite patient safety score for U.S. hospitals. *Journal of Patient Safety* 10(1):64–71.

Austin PC. (2002) A comparison of Bayesian methods for profiling hospital performance. *Medical Decision Making* 22:163–172.

Austin PC. (2009) The relative ability of different propensity score methods to balance measured covariates between treated and untreated subjects in observational studies. *Medical Decision Making* 29(6):661–677.

Austin PC. (2014) A comparison of 12 algorithms for matching on the propensity score. *Statistics in Medicine* 33(6):1057–1069.

Austin PC, Alter DA, Anderson GM, Tu JV. (2004) Impact of the choice of benchmark on the conclusions of hospital report cards. *American Heart Journal* 148:1041–1046.

Austin PC, Alter DA, Tu JV. (2003) The use of fixed-and random-effects models for classifying hospitals as mortality outliers: A Monte Carlo assessment. *Medical Decision Making* 23:526–539.

Austin PC, Reeves MJ. (2013) The relationship between the C-statistic of a risk-adjustment model and the accuracy of hospital report cards: A Monte Carlo Study. *Medical Care* 51(3):275–284.

Austin PC, Tu JV. (2004) Automated variable selection methods for logistic regression produced unstable models for predicting acute myocardial infarction mortality. *Journal of Clinical Epidemiology* 57(11):1138–1146.

Austin PC, Tu JV. (2006) Comparing clinical data with administrative data for producing acute myocardial infarction report cards. *Journal of the Royal Statistical Society Series A* 169(1):115–126.

Australian Council for Safety and Quality in Health Care (July 2002) Lessons from the inquiry into obstetrics and gynaecological services at King Edward Memorial Hospital 1990–2000. Available at http://www.safetyandquality.gov.au/wp-content/uploads/2012/01/king_edward.pdf (accessed September 2015).

Australian Institute of Health and Welfare. (2008) A set of performance indicators across the health and aged care system for Health Ministers. Available at http://www.aihw.gov.au/WorkArea/DownloadAsset.aspx?id=6442471955 (accessed September 2015).

References

Aylin P, Alves B, Best N, Cook A, Elliott P, Evans SJ, Lawrence AE, Murray GD, Pollock J, Spiegelhalter D. (2001) Comparison of UK paediatric cardiac surgical performance by analysis of routinely collected data 1984–96: Was Bristol an outlier? *Lancet* 358(9277):181–187.

Aylin P, Alves B, Cook A, Bennett J, Bottle A, Best N, Catena B, Elliott P. (1999) Analysis of hospital episode statistics for the Bristol royal infirmary inquiry, Division Primary Care & Population Health Sciences, Imperial College School of Medicine, London, UK, October 1999.

Aylin P, Best N, Bottle A, Marshall C. (2003) Following Shipman: A pilot system for monitoring mortality rates in primary care. *Lancet* 362(9382):485–491.

Aylin P, Bottle A, Majeed A. (2007b) Use of administrative data or clinical databases as predictors of risk of death in hospital: Comparison of models. *British Medical Journal* 334:1044.

Aylin P, Lees T, Baker S, Prytherch D, Ashley S. (2007a) Descriptive study comparing routine hospital administrative data with the Vascular Society of Great Britain and Ireland's National Vascular Database. *European Journal of Vascular and Endovascular Surgery* 33(4):461–465.

Bahl V, Thompson MA, Kau TY, Hu HM, Campbell DA Jr. (2008) Do the AHRQ patient safety indicators flag conditions that are present at the time of hospital admission? *Medical Care* 46(5):516–522.

Ballentine NH. (2009) Coding and documentation: Medicare severity diagnosis-related groups and present-on-admission documentation. *Journal of Hospital Medicine* 4:124–130.

Bardsley M, Spiegelhalter DJ, Blunt I, Chitnis X, Roberts A, Bharania S. (2009) Using routine intelligence to target inspection of healthcare providers in England. *Quality and Safety in Health Care* 18(3):189–194.

Barton JL, Imboden J, Graf J, Glidden D, Yelin EH, Schillinger D. (2010) Patient-physician discordance in assessments of global disease severity in rheumatoid arthritis. *Arthritis Care Research (Hoboken)* 62(6):857–864.

Beaulieu N, Epstein AM. (2002) National committee on quality Assurance health-plan accreditation: Predictors, correlates of performance, and market impact. *Medical Care* 40:325–337.

Benjamini Y, Hochberg Y. (1995) Controlling the false discovery rate: A practical and powerful approach to multiple testing. *Journal of the Royal Statistical Society Series B* 57:289–300.

Berwick D. (2013) *A Promise to Learn – A Commitment to Act: Improving the Safety of Patients in England*. London, UK: National Advisory Group on the Safety of Patients in England.

Berwick DM, Godfrey AB, Roessner J. (2002) *Curing Health Care: New Strategies for Quality Improvement*. San Francisco, CA: Jossey-Bass.

Bevan G, Hood C. (2006) Have targets improved performance in the English NHS? *British Medical Journal* 332:419.

Black N, Barker M, Payne M. (2004) Cross-sectional survey of multicentre clinical databases in the United Kingdom. *British Medical Journal* 328(7454):1478.

Bland JM, Altman DG. (1994) Statistics notes: Correlation, regression, and repeated data. *British Medical Journal* 308:896.

Blozik E, Nothacker M, Bunk T, Szecsenyi J, Ollenschläger G, Scherer M. (2012) Simultaneous development of guidelines and quality indicators – How do guideline groups act? A worldwide survey. *International Journal of Health Care Quality Assurance* 25(8):712–729.

Boaden R, Harvey G, Moxham C, Proudlove N. (2008) *Quality Improvement: Theory and Practice in Healthcare*. Coventry, UK: NHS Institute for Innovation and Improvement.

Board N, Matete-Owiti S. Review of patient experience and satisfaction surveys conducted within public and private hospitals in Australia. Australian Commission on Safety and Quality in Health Care 2012. Available at http://www.safetyandquality.gov.au/wp-content/uploads/2012/03/Review-of-Hospital-Patient-Experience-Surveys-conducted-by-Australian-Hospitals-30-March-2012-FINAL.pdf (accessed May 2015).

Bobko P, Roth PL, Buster MA. (2007) The usefulness of unit weights in creating composite scores: A literature review, application to content validity, and meta-analysis. *Organizational Research Methods* 10:689–709.

Bottle A, Aylin P. (2008a) Intelligent information: A national system for monitoring clinical performance. *Health Services Research* 43(1 Pt 1):10–31.

Bottle A, Aylin P. (2008b) How NHS trusts could use patient safety indicators to help improve care. *Health Care Risk Report*, May 12–14.

Bottle A, Aylin P. (2009) Application of AHRQ patient safety indicators to English hospital data. *Quality and Safety in Health Care* 18(4):303–308.

Bottle A, Aylin P. (2011) Predicting the false alarm rate in multi-institution mortality monitoring. *Journal of the Operational Research Society* 62:1711–1718.

Bottle A, Gaudoin R, Goudie R, Jones S, Aylin P. (2014) Can valid and practical risk-prediction or casemix adjustment models, including adjustment for comorbidity, be generated from English hospital administrative data (Hospital Episode Statistics)? A National Observational Study. Southampton, UK: NIHR Journals Library, November.

Bottle A, Jarman B, Aylin P. (2010) Strengths and weaknesses of hospital standardised mortality ratios. *British Medical Journal* 342:c7116.

Bottle A, Jarman B, Aylin P. (2011) Hospital standardized mortality ratios: Sensitivity analyses on the impact of coding. *Health Services Research* 46(6pt1):1741–1761.

Bottle A, Middleton S, Kalkman CJ, Livingston EH, Aylin P. (2013a) Global comparators project: International comparison of hospital outcomes using administrative data. *Health Service Research* 48(6 Pt 1):2081–2100.

Bottle A, Sanders RD, Mozid A, Aylin P. (2013b) Provider profiling models for acute coronary syndrome mortality using administrative data. *International Journal of Cardiology* 168(1):338–343.

Bourn J. (2007) *Prescribing Costs in Primary Care*. London, UK: National Audit Office.

Bowling A, Ebrahim S. (2001) Measuring patients' preferences for treatment and perceptions of risk. *Quality in Health Care* 10:i2–i8.

Brennan TA, Berwick DM. (1996) *New Rules: Regulation, Markets, and the Quality of American Health Care*. San Francisco, CA: Jossey-Bass.

Brennan TA, Leape LL, Laird NM, Hebert L, Localio AR, Lawthers AG, Newhouse JP, Weiler PC, Hiatt HH. (1991) Incidence of adverse events and negligence in hospitalized patients. Results of the Harvard Medical Practice Study I. *New England Journal of Medicine* 324(6):370–376.

Brent RJ. (2003) *Cost-Benefit Analysis and Health Care Evaluations*. Cheltenham, UK: Edward Elgar Publishing.

Breslow NE, Day NE. (1987) *Statistical Methods in Cancer Research, Volume II: The Design and Analysis of Cohort Studies*. New York: Oxford University Press.

Bunce C, Patel KV, Xing W, Freemantle N, Doré CJ on behalf of the Ophthalmic Statistics Group. (2014) Ophthalmic statistics note 1: Unit of analysis. *British Journal of Ophthalmology* 98:408–412.

Burns EM, Rigby E, Mamidanna R, Bottle A, Aylin P, Ziprin P, Faiz OD. (2012) Systematic review of discharge coding accuracy. *Journal of Public Health (Oxford)* 34(1):138–148.

Cameron AC, Miller DL. (2015) A practitioner's guide to cluster-robust inference. *Journal of Human Resources* 50(2):317–372.

Cantillon P, Sargent J. (2008) Giving feedback in clinical settings. *British Medical Journal* 337:a1961.

Carey RG. (2003) *Improving Healthcare with Control Charts. Basic and Advanced SPC Methods and Case Studies*. Milwaukee, WI: American Society for Quality.

Carrière I, Bouyer J. (2002) Choosing marginal or random-effects models for longitudinal binary responses: Application to self-reported disability among older persons. *BMC Medical Research Methodology* 2:15.

Charlson ME, Pompei P, Ales KL, MacKenzie CR. (1987) A new method of classifying prognostic comorbidity in longitudinal studies: Development and validation. *Journal of Chronic Disease* 40(5):373–383.

Chen LM, Staiger DO, Birkmeyer JD, Ryan AM, Zhang W, Dimick JB. (2013) Composite quality measures for common inpatient medical conditions. *Medical Care* 51:832–837.

Chong CAKY, Nguyen GC, Wilcox ME. (2012) Trends in Canadian hospital standardised mortality ratios and palliative care coding 2004–2010: A retrospective database analysis. *British Medical Journal Open* 2:e001729.

Christiansen CL, Morris CN. (1997) Improving the statistical approach to health care provider profiling. *Annals of Internal Medicine* 127:764–768.

Cima RR, Lackore KA, Nehring SA, Cassivi SD, Donohue JH, Deschamps C, Vansuch M, Naessens JM. (2011) How best to measure surgical quality? Comparison of the Agency for Healthcare Research and Quality Patient Safety Indicators (AHRQ-PSI) and the American College of Surgeons National Surgical Quality Improvement Program (ACS-NSQIP) postoperative adverse events at a single institution. *Surgery* 150(5):943–949.

Clarke P. (2008) When can group level clustering be ignored? Multilevel models versus single-level models with sparse data. *Journal of Epidemiology and Community Health* 62(8):752–758.

Clarke S, Oakley J (eds). (2007) *Informed Consent and Clinician Accountability: The Ethics of Report Cards on Surgeon Performance*. Cambridge, UK: Cambridge University Press.

Cohen ME, Ko CY, Bilimoria KY, Zhou L, Huffman K, Wang X, Liu Y et al. (2013) Optimizing ACS NSQIP modeling for evaluation of surgical quality and risk: Patient risk adjustment, procedure mix adjustment, shrinkage adjustment, and surgical focus. *Journal of the American College of Surgeons* 217:336–346.

Colla JB, Bracken AC, Kinney LM, Weeks WB. (2005) Measuring patient safety climate: A review of surveys. *Quality and Safety in Health Care* 14:364–366.

Collopy BT, Williams J, Rodgers L, Campbell J, Jenner N, Andrews N. (2000) The ACHS care evaluation program: A decade of achievement. Australian Council on Healthcare Standards. *Journal of Quality in Clinical Practice* 20(1):36–41.

Commission for Health Improvement. (2003) *What CHI Has Found in Ambulance Trusts*. London, UK: Stationery Office.

Conover, C. (2004) Health care regulation. Policy Analysis no. 527.

Cook N. (2007) Use and misuse of the receiver operating characteristic curve in risk prediction. *Circulation* 115:928–935.

Copas JB. (1983) Regression, prediction and shrinkage. *Journal of the Royal Statistical Society Series B* 45(3):311–354.

Corrigan JM, Donaldson MS, Kohn LT (eds). (2001) *Crossing the Quality Chasm: A New Health System for the 21st Century*. Washington, DC: National Academy.

Craig P, Cooper C, Gunnell D, Haw S, Lawson K, Macintyre S, Ogilvie D et al. (2012) Using natural experiments to evaluate population health interventions: New Medical Research Council guidance. *Journal of Epidemiology and Community Health* 66:1182–1186.

Cruz JA, Wishart DS. (2006) Applications of machine learning in cancer prediction and prognosis. *Cancer Informatics* 2:59–77.

Curry LA, Nembhard IM, Bradley EH. (2009) Qualitative and mixed methods provide unique contributions to outcomes research. *Circulation* 119:1442–1452.

D'Lima DM, Moore J, Bottle A, Brett SJ, Arnold GM, Benn J. (2015) Developing effective feedback on quality of anaesthetic care: What are its most valuable characteristics from a clinical perspective? *Journal of Health Services Research & Policy* 20(S1):26–34.

Damberg CL, Sorbero ME, Lovejoy SL, Lauderdale K, Wertheimer S, Smith A, Waxman D, Schnyer C. (2012) An evaluation of the use of performance measures in health care. *RAND Health Quarterly* 1(4):3.

Damman OC, van den Hengel YK, van Loon AJM, Rademakers J. (2010) An international comparison of web-based reporting about health care quality: Content analysis. *Journal of Medical Internet Research* 12:e8.

De Silva D. 2013. Measuring patient experience. Health Foundation. Available at http://www.health.org.uk/publication/measuring-patient-experience (accessed September 2015).

de Vos M, Graafmans W, Kooistra M, Meijboom B, van der Voort P, Westert G. (2009) Using quality indicators to improve hospital care: A review of the literature. *International Journal for Quality in Health Care* 21(2):119–129.

Degenholtz HB, Bhatnagar M. (2009) Introduction to hierarchical modeling. *Journal of Palliative Medicine* 12(7):631–638.

DeLong ER, Peterson ED, Delong DM, Muhlbaier LH, Hackett S, Mark DB. (1997) Comparing risk-adjustment methods for provider profiling. *Statistics in Medicine* 16:2645–2664.

Devkaran S, O'Farrell PN. (2014) The impact of hospital accreditation on clinical documentation compliance: A life cycle explanation using interrupted time series analysis. *British Medical Journal Open* 4:e005240.

Dhalla IA, O'Brien T, Morra D, Thorpe KE, Wong BM, Mehta R, Frost DW et al. (2014) Effect of a postdischarge virtual ward on readmission or death for high-risk patients: A randomized clinical trial. *Journal of the American Medical Association* 312(13):1305–1312.

Diehr P, Cain K, Connell F, Volinn E. (1990) What is too much variation? The null hypothesis in small-area analysis. *Health Services Research* 24(6):741–771.

Dimick JB, Ghaferi AA, Osborne NH, Ko CY, Hall BL. (2012) Reliability adjustment for reporting hospital outcomes with surgery. *Annals of Surgery* 255(4):703–707.

Dimick JB, Hendren SK. (2014) Hospital report cards: Necessary but not sufficient? *Journal of the American Medical Association Surgery* 149(2):143–144.

Dimick JB, Welch HG, Birkmeyer JD. (2004) Surgical mortality as an indicator of hospital quality: The problem with small sample size. *Journal of the American Medical Association* 292(7):847–851.

Dixon A, Robertson R, Appleby J, Burge P, Devlin N, Magee H. (2010) *Patient Choice. How Patients Choose and How Providers Respond*. London, UK: King's Fund.

Donabedian A. (1966) Evaluating the quality of medical care. *Milbank Memorial Fund Quarterly* 44:166–206.

Donders AR, van der Heijden GJ, Stijnen T, Moons KG. (2006) Review: A gentle introduction to imputation of missing values. *Journal of Clinical Epidemiology* 59:1087–1091.

Dorsey D, Grady JN, Desai N, Lindenauer PM, Bernheim S, Young J, Zhang W et al. (2015) 2015 Condition-specific measures updates and specifications report. Hospital-level 30-day risk-standardized mortality measures. YNHHSC/CORE, March.

Dosh SA, Hickner JM, Mainous AG 3rd, Ebell MH. (2000) Predictors of antibiotic prescribing for nonspecific upper respiratory infections, acute bronchitis, and acute sinusitis. An UPRNet study. Upper Peninsula Research Network. *Journal of Family Practice* 49(5):407–414.

Dreiher J, Comaneshter DS, Rosenbluth Y, Battat E, Bitterman H, Cohen AD. (2012) The association between continuity of care in the community and health outcomes: A population based study. *Israel Journal of Health Policy Research* 1:21.

Dreiseitl S, Ohno-Machado L. (2002) Logistic regression and artificial neural network classification models: A methodology review. *Journal of Biomedical Informatics* 35:352–359.

Drummond MF, Sculpher MJ, Torrance GW, O'Brien BJ, Stoddart GL. (2005) *Methods for the Economic Evaluation of Health Care Programme*, 3rd edn. New York: Oxford University Press.

Eccles M, Fremantle N, Mason J. (1998) North of England evidence based guidelines development project: Methods of developing guidelines for efficient drug use in primary care. *British Medical Journal* 316:1232–1235.

Eftekar B, Mohammad K, Eftekhar Ardebili H, Ghodsi M, Ketabchi E. (2005) Comparison of artificial neural network and logistic regression models for prediction of mortality in head trauma based on initial clinical data. *BMC Medical Informatics and Decision Making* 5:3.

Elixhauser A, Steiner C, Harris DR, Coffey RM. (1998) Comorbidity measures for use with administrative data. *Medical Care* 36(1):8–27.

Elliott MN, Zaslavsky AM, Goldstein E, Lehrman W, Hambarsoomians K, Beckett MK, Giordano L. (2009) Effects of survey mode, patient mix, and nonresponse on CAHPS hospital survey scores. *Health Services Research* 44(2 Pt 1):501–518.

European Collaboration for Healthcare Optimization (ECHO) Project. www.echohealth.eu. Zaragoza (Spain): Instituto Aragonés de Ciencias de la Salud – Instituto Investigación Sanitaria Aragón. (c2010) Estupiñán Romero FR, Baixauli Pérez C, Bernal-Delgado E on behalf of the ECHO consortium. Handbook on methodology: ECHO information system quality report, April 2014. Available at www.echo-health.eu/echo-atlas-reports (accessed March 2015).

European Commission. (2014) Patient Safety and Quality of Care working group. Key findings and recommendations on reporting and learning systems for patient safety incidents across Europe. Report of the reporting and learning subgroup of the European Commission, May. Available at http://ec.europa.eu/health/patient_safety/policy/index_en.htm (accessed on 4 April 2016).

Evans SM, Berry JG, Smith BJ, Esterman A, Selim P, O'Shaughnessy J, DeWit M. (2006) Attitudes and barriers to incident reporting: A collaborative hospital study. *Quality and Safety in Health Care* 15(1):39–43.

Fabri PJ. (2014) Classification techniques in analyzing surgical outcomes data. *Journal of the American College of Surgeons* 218(2):283–289.

Falstie-Jensen AM, Larsson H, Hollnagel E, Nørgaard M, Svendsen MLO, Johnsen SP. (2015) Compliance with hospital accreditation and patient mortality: A Danish nationwide population-based study. *International Journal of Quality in Health Care* 27(3):165–174.

Fan VS, Au D, Heagerty P, Deyo RA, McDonell MB, Fihn SD. (2002) Validation of case-mix measures derived from self-reports of diagnosis and health. *Journal of Clinical Epidemiology* 55(4):371–380.

Farley DO, Haviland A, Champagne S, Jain AK, Battles JB, Munier WB, Loeb JM. (2008) Adverse-event-reporting practices by US hospitals: Results of a national survey. *Quality and Safety in Health Care* 17:416–423.

Fedeli U, Broccoa S, Alba N, Rosato R, Spolaore P. (2007) The choice between different statistical approaches to risk-adjustment influenced the identification of outliers. *Journal of Clinical Epidemiology* 60:858–862.

Ferreira-Gonzalez I, Marsal JR, Mitjavila F, Parada A, Ribera A, Cascant P, Soriano N et al. (2009) Patient registries of acute coronary syndrome: Assessing or biasing the clinical real world data? *Circulation: Cardiovascular Quality and Outcomes* 2:540–547.

Fisher ES, Wennberg DE, Stukel TA, Gottlieb DJ, Lucas FL, Pinder EL. (2003) The implications of regional variations in Medicare spending, Part 2: Health outcomes and satisfaction with care. *Annals of Internal Medicine* 138:288–298.

Fitzmaurice G, Laird N, Ware J. (eds). (2004) *Applied Longitudinal Analysis*. Hoboken, NJ: Wiley, pp. 87–102.

Flodgren G, Pomey MP, Taber SA, Eccles MP. (2011) Effectiveness of external inspection of compliance with standards in improving healthcare organisation behaviour, healthcare professional behaviour or patient outcomes. *Cochrane Database of Systematic Reviews* 11:CD008992.

Forrey RA, Pedersen CA, Schneider PJ. (2007) Interrater agreement with a standard scheme for classifying medication errors. *American Journal of Health-System Pharmacy* 64(2):175–181.

Fox KAA, Goodman SG, Klein W, Brieger D, Steg PG, Dabbous O, Avezum Á for the GRACE Investigators. (2002) Management of acute coronary syndromes. Variations in practice and outcome: Findings from the Global Registry of Acute Coronary Events (GRACE). *European Heart Journal* 23:1177–1189.

Fox S. (2011) *The Social Life of Health Information*. Washington, DC: Pew Research Center. Available at http://www.pewinternet.org/2011/05/12/the-social-life-of-health-information-2011/ (accessed September 2015).

Francis R. (2013) *Report of the Mid Staffordshire NHS Foundation Trust Public Inquiry*. London, UK: The Stationery Office.

Fresko A, Rubenstein S. (2013) Taking safety on board: The board's role in patient safety. Health Foundation, October. Available at http://www.health.org.uk/sites/default/files/TakingSafetyOnBoardTheBoardsRoleInPatientSafety.pdf (accessed September 2015).

Frisen M. (1992) Evaluations of methods for statistical surveillance. *Statistics in Medicine* 11:1489–1502.

Frisen M. (2003) Statistical surveillance. Optimality and methods. *International Statistical Review* 71(2):403–434.

Fung CH, Lim YW, Mattke S, Damberg C, Shekelle PG. (2008) Systematic review: The evidence that publishing patient care performance data improves the quality of care. *Annals of Internal Medicine* 148:111–123.

Garout M, Tilney HS, Tekkis PP, Aylin P. (2008) Comparison of administrative data with the Association of Coloproctology of Great Britain and Ireland (ACPGBI) colorectal cancer database. *International Journal of Colorectal Disease* 23(2):155–163.

Genell A, Nemes S, Steineck G, Dickman PW. (2010) Model selection in medical research: A simulation study comparing Bayesian model averaging and stepwise regression. *BMC Medical Research Methodology* 10:108.

Geraci JM, Johnson ML, Gordon HS, Petersen NJ, Shroyer AL, Grover FL, Wray NP. (2005) Mortality after cardiac bypass surgery prediction from administrative versus clinical data. *Medical Care* 43:149–158.

Geraedts M, Hermeling P, de Cruppé W. (2012) Communicating quality of care information to physicians: a study of eight presentation formats. *Patient Education and Counselling* 87(3):375–382.

Gerteis M, Gerteis JS, Newman D, Koepke C. (2007) Testing consumers' comprehension of quality measures using alternative reporting formats. *Health Care Finance Review* 28:31–45.

Glance LG, Dick AW, Osler TM, Mukamel D. (2003) Using hierarchical modeling to measure ICU quality. *Intensive Care Medicine* 29:2223–2229.

Glance LG, Osler TM, Mukamel DB, Meredith W, Dick AW. (2009) Impact of statistical approaches for handling missing data on trauma center quality. *Annals of Surgery* 249:143–148.

Glance LG, Osler TM, Mukamel DB, Meredith JW, Dick AW. (2014) Effectiveness of nonpublic report cards for reducing trauma mortality. *Journal of the American Medical Association Surgery* 149(2):137–143.

Gliklich RE, Dreyer NA (eds). (2010) *Registries for Evaluating Patient Outcomes: A User's Guide*, 2nd edn. AHRQ Publication No. 10-EHC049. Rockville, MD: Agency for Healthcare Research and Quality, September.

Goldman LE, Chu PW, Osmond D, Bindman A. (2011) The accuracy of present-on-admission reporting in administrative data. *Health Services Research* 46 (6 pt 1):1946–1962.

Goldstein HJ, Spiegelhalter DJ. (1996) League tables and their limitations: Statistical issues in comparisons of institutional performance. *Journal of the Royal Statistical Society Series A* 159(3):385–443.

Gordon HS, Johnson ML, Wray NP, Petersen NJ, Henderson WG, Khuri SF, Geraci JM. (2005) Mortality after noncardiac surgery prediction from administrative versus clinical data. *Medical Care* 43:159–167.

Greaves F, Laverty AA, Cano DR, Moilanen K, Pulman S, Darzi A, Millett C. (2014) Tweets about hospital quality: A mixed methods study. *British Medical Journal Quality and Safety* 23(10):838–846.

Greaves F, Ramirez-Cano D, Millett C, Darzi A, Donaldson L. (2013) Use of sentiment analysis for capturing patient experience from free-text comments posted online. *Journal of Medical Internet Research* 15(11):e239.

Green J, Wintfeld N. (1995) Report cards on cardiac surgeons. Assessing New York State's approach. *New England Journal of Medicine* 332:1229–1232.

Greenfield D, Braithwaite J. (2008) Health sector accreditation research: A systematic review. *International Journal of Quality in Health Care* 20(3):172–183.

Greenland S. (2000) Principles of multilevel modelling. *International Journal of Epidemiology* 29:158–167.

Greenland S, Robins JM. (1991) Empirical Bayes adjustments for multiple comparisons are sometimes useful. *Epidemiology* 2(4):244–251.

Guthrie B, Love T, Fahey T, Morris A, Sullivan F. (2005) Control, compare and communicate: Designing control charts to summarise efficiently data from multiple quality indicators. *Quality and Safety in Health Care* 14:450–454.

Haas J, Luft H, Romano PS, Dean M, Hung Y, Bacchetti P. (2010) Report for the California Hospital Outcomes Project. Community-acquired pneumonia, 1996: Model Development and Validation. Office of Statewide Health Planning and Development, Sacramento, CA.

Ham C, Raleigh V, Foot C, Robertson R, Alderwick H. (2015) Measuring the performance of local health systems: A review for the Department of Health. The King's Fund, London, UK.

Hammill BG, Curtis LH, Fonarow GC, Heidenreich PA, Yancy CW, Peterson ED, Hernandez AF. (2011) Incremental value of clinical data beyond claims data in predicting 30-day outcomes after heart failure hospitalization. *Circulation: Cardiovascular Quality and Outcomes* 4:60–67.

Hannan EL, Kilburn H, Racz M, Shields E, Chassin MR. (1994) Improving the outcomes of coronary artery bypass surgery in New York State. *Journal of the American Medical Association* 271:761–766.

Harman JS, Scholle SH, Ng JH, Pawlson LG, Mardon RE, Haffer SC, Shih S, Bierman AS. (2010) Association of Health Plans' Healthcare Effectiveness Data and Information Set (HEDIS) performance with outcomes of enrollees with diabetes. *Medical Care* 48(3):217–223.

Harrell FE Jr, Lee KL, Califf RM, Pryor DB, Rosati RA. (1984) Regression modelling strategies for improved prognostic prediction. *Statistics in Medicine* 3(2):143–152.

Hauck K, Rice N, Smith P. (2003) The influence of health care organisations on health system performance. *Journal of Health Services Research and Policy* 8(2):68–74.

Hayward RA. (2007) All-or-nothing treatment targets make bad performance measures. *American Journal of Managed Care* 13:126–128.

Hayward RA, Hofer TP. (2001) Estimating hospital deaths due to medical errors: Preventability is in the eye of the reviewer. *Journal of the American Medical Association* 286(4):415–420.

Health and Social Care Information Centre. (2013) The use of palliative care coding in the summary hospital-level mortality indicator. Available at http://www.hscic.gov.uk/media/11150/Palliative-Care-Coding-Report/pdf/Palliative_Care_Coding_Report.pdf (accessed September 2015).

… *References* … 247

Health Protection Agency. (2012) *English National Point Prevalence Survey on Healthcare-associated Infections and Antimicrobial Use, 2011*. London, UK: Health Protection Agency.

Herrett E, Shah AD, Boggon R, Denaxas S, Smeeth L, van Staa T, Timmis A, Hemingway H. (2013) Completeness and diagnostic validity of recording acute myocardial infarction events in primary care, hospital care, disease registry, and national mortality records: Cohort study. *British Medical Journal* 346:f2350.

Hesketh EA, Laidlaw JM. (2002) Developing the teaching instinct: 1: Feedback. *Medical Teacher* 24:245–248.

Hibbard JH, Peters E. (2003) Supporting informed consumer health care decisions: Data presentation approaches that facilitate the use of information in choice. *Annual Review of Public Health* 24:413–433.

Hibbard JH, Stockard J, Tusler M. (2003) Does publicizing hospital performance stimulate quality improvement efforts? *Health Affairs* 22(2):84–94.

Hickey GL, Grant SW, Caiado C, Kendall S, Dunning J, Poullis M, Buchan I, Bridgewater B. (2013b) Dynamic prediction modeling approaches for cardiac surgery. *Circulation: Cardiovascular Quality Outcomes* 6:649–658.

Hickey GL, Grant SW, Murphy GJ, Bhabra M, Pagano D, McAllister K. (2013a) Dynamic trends in cardiac surgery: Why the logistic EuroSCORE is no longer suitable for contemporary cardiac surgery and implications for future risk models. *European Journal Cardio-Thoracic Surgery* 43:1146–1152.

Hoeting JA, Madigan D, Raftery AE, Volinsky CT. (1999) Bayesian model averaging: A tutorial. *Statistical Science* 14(4):382–417.

Hofer TP, Hayward RA. (1996) Identifying poor-quality hospitals. Can hospital mortality rates detect quality problems for medical diagnoses? *Medical Care* 34(8):737–753.

Hofer TP, Hayward RA, Greenfield S, Wagner EH, Kaplan SH, Manning WG. (1999) The unreliability of individual physician "Report Cards" for assessing the costs and quality of care of a chronic disease. *Journal of the American Medical Association* 281(22):2098–2105.

Hollingsworth B. (2003) Non-parametric and parametric applications measuring efficiency in health care. *Health Care Management Science* 6:203–218.

Hornbrook MC, Hurtado AV, Johnson RE. (1985) Health care episodes: Definition, measurement and use. *Medical Care Reviews* 42(2):163–218.

Hosmer DW, Lemeshow S. (1989) *Applied Logistic Regression*. New York: John Wiley.

Hosmer DW, Lemeshow S. (1995) Confidence interval estimates of an index of quality performance based on logistic regression models. *Statistics in Medicine* 14(19):2161–2172.

Hsee CK, Zhang J. (2010) General evaluability theory. *Perspectives on Psychological Science* 5(4):343–355.

Hu B, Palta M, Shao J. (2006) Properties of R^2 statistics for logistic regression. *Statistics in Medicine* 25:1383–1395.

Hu FB, Goldberg J, Hedeker D, Flay BR, Pentz MA. (1998) Comparison of population-averaged and subject-specific approaches for analysing repeated binary outcomes. *American Journal of Epidemiology* 147(7):694–703.

Hubbard AE, Ahern J, Fleischer NL, Van der Laan M, Lippman SA, Jewell N, Bruckner T, Satariano WA. (2010) To GEE or Not to GEE. Comparing population average and mixed models for estimating the associations between neighborhood risk factors and health. *Epidemiology* 21:467–474.

Hurst J, Jee-Hughes M. (2001) *Performance Measurement and Performance Management in OECD Health Systems*. OECD Labour Market and Social Policy Occasional Papers, No. 47, Paris, France: OECD Publishing. Available at http://dx.doi.org/10.1787/788224073713 (accessed September 2015).

Hussey PS, de Vries H, Romley J, Wang MC, Chen SS, Shekelle PG, McGlynn EA. (2009) A systematic review of health care efficiency measures. *Health Services Research* 44(3):784–805.

Hutchinson A, Coster JE, Cooper KL, McIntosh A, Walters SJ, Bath PA, Pearson M et al. (2010) Comparison of case note review methods for evaluating quality and safety in health care. *Health Technology Assessment* 14(10):1–144.

Hutchinson A, Coster JE, Cooper KL, Pearson M, McIntosh A, Bath PA. (2013) A structured judgement method to enhance mortality case note review: Development and evaluation. *British Medical Journal Quality and Safety* 22(12):1032–1040.

Ibáñez B, Librero J, Bernal-Delgado E, Peiró S, González López-Valcarcel B, Martínez N, Aizpuru F. (2009) Is there much variation in variation? Revisiting statistics of small area variation in health services research. *BMC Health Services Research* 9:60.

Iezzoni LI. (2013) *Risk Adjustment for Measuring Healthcare Outcomes*, 4th edn. Chicago, IL: Health Administration Press.

Iezzoni LI, Davis RB, Palmer RH, Cahalane M, Hamel MB, Mukamal K, Phillips RS, Banks NJ, Davis DT Jr. (1999) Does the complications screening program flag cases with process of care problems? Using explicit criteria to judge processes. *International Journal of Quality in Health Care* 11(2):107–118.

Imbens GW, Lemieux T. (2008) Regression discontinuity designs: A guide to practice. *Journal of Econometrics* 142:615–635.

Institute of Medicine. (2000) *To Err Is Human: Building a Safer Health System*. Kohn LT, Corrigan JM, Donaldson MS (eds). Washington, DC: National Academy Press.

Ivers N, Jamtvedt G, Flottorp S, Young JM, Odgaard-Jensen J, French SD, O'Brien MA, Johansen M, Grimshaw J, Oxman AD. (2012) Audit and feedback: Effects on professional practice and healthcare outcomes. *Cochrane Database of Systematic Reviews* 6:CD000259.

Jacobs R. (2000) Alternative methods to examine hospital efficiency: Data envelopment analysis and stochastic frontier analysis. Centre for Health Economics Discussion Paper 177. University of York. Available at http://www.york.ac.uk/che/pdf/DP177.pdf (accessed September 2015).

Jacobs R, Smith P, Goddard M. (2004) Measuring performance: An examination of composite performance indicators. CHE Technical Paper Series 29. University of York, York, UK. Available at http://www.york.ac.uk/che/pdf/tp29.pdf (accessed September 2015).

Jacobs R, Smith P, Street A. (2006) Measuring efficiency in health care: analytic techniques and health policy. Cambridge, UK: Cambridge University Press.

Januel J-M, Luthi J-C, Quan H, Borst F, Taffé P, Ghali WA, Burnand B. (2011) Improved accuracy of co-morbidity coding over time after the introduction of ICD-10 administrative data. *BMC Health Services Research* 11:194.

Jarman B, Gault S, Alves B, Hider A, Dolan S, Cook A, Hurwitz B, Iezzoni LI. (1999) Explaining differences in English hospital death rates using routinely collected data. *British Medical Journal* 318(7197):1515–1520.

Jen MH, Bottle A, Kirkwood G, Johnston R, Aylin P. (2011) The performance of automated case-mix adjustment regression model building methods in a health outcome prediction setting. *Health Care Management Science* 14(3):267–278.

Jha AK, DesRoches CM, Kralovec PD, Joshi MS. (2010) A progress report on electronic health records in US Hospitals. *Health Affairs* 29(10):1951–1957.
Jha AK, Zaslavsky AM. (2014) Quality reporting that addresses disparities in health care. *Journal of the American Medical Association* 312(3):225–226.
Johnston MV, Ottenbacher KJ, Reichardt CS. (1995) Strong quasi-experimental designs for research on the effectiveness of rehabilitation. *American Journal of Physical Medicine & Rehabilitation* 74(5):383–392.
Jones C, Gannon B, Wakai A, O'Sullivan R. (2015) A systematic review of the cost of data collection for performance monitoring in hospitals. *Systematic Reviews* 4:38.
Jones HE, Ohlssen DI, Spiegelhalter DJ. (2008) Use of the false discovery rate when comparing multiple health care providers. *Journal of Clinical Epidemiology* 61:232–240.
Jones HE, Spiegelhalter DJ. (2011) The identification of "Unusual" health-care providers from a hierarchical model. *American Statistician* 65(3):154–163.
Julious SA, Nicholl J, George S. (2001) Why do we continue to use standardized mortality ratios for small area comparisons? *Journal of Public Health Medicine* 23(1):40–46.
Kahn JM, Kramer AA, Rubenfeld GD. (2007) Transferring critically ill patients out of hospital improves the standardized mortality ratio. *Chest* 131:68–75.
Kalbfleisch JD, Wolfe RA. (2013) On monitoring outcomes of medical providers. *Statistics in Biosciences* 5:286–302.
Kammermeister K. (2009) The national surgical quality improvement program: Learning from the past and moving to the future. *American Journal of Surgery* 198(5 Suppl):S69–S73.
Kao LS, Ghaferi AA, Ko CY, Dimick JB. (2011) Reliability of superficial surgical site infections as a hospital quality measure. *Journal of the American College Surgery* 213:231–235.
Kaplan RS, Norton DP. (1996) *The Balanced Scorecard: Translating Strategy into Action*. Boston, MA: Harvard Business School Press.
Kaplan SH, Griffith JL, Price LL, Pawlson LG, Greenfield S. (2009) Improving the reliability of physician performance assessment: Identifying the "Physician Effect" on quality and creating composite measures. *Medical Care* 47:378–387.
Kawachi I, Subramanian SV, Almeida-Filho N. (2002) A glossary for health inequalities. *Journal of Epidemiology and Community Health* 56:647–652.
Kaye AD, Okanlawon OJ, Urman RD. (2014) Clinical performance feedback and quality improvement opportunities for perioperative physicians. *Advances in Medical Education and Practice* 5:115–123.
Kennedy I. (2001) *The Report of the Public Inquiry into Children's Heart Surgery at the Bristol Royal Infirmary 1984–1995: Learning from Bristol*. London, UK: The Stationery Office.
Keogh B. (2013) Review into the quality of care and treatment provided by 14 hospital trusts in England: Overview report. NHS England. Available at http://www.nhs.uk/nhsengland/bruce-keogh-review/documents/outcomes/keogh-review-final-report.pdf (accessed October 2015).
Kerry SM, Bland JM. (1998) Analysis of a trial randomised in clusters. *British Medical Journal* 316(7124):54.
Ketelaar NABM, Faber MJ, Flottorp S, Rygh LH, Deane KHO, Eccles MP. (2011) Public release of performance data in changing the behaviour of healthcare consumers, professionals or organisations. *Cochrane Library* 11:CD004538.
Klabunde CN, Reeve BB, Harlan LC, Davis WW, Potosky AL. (2005) Do patients consistently report comorbid conditions over time? Results from the prostate cancer outcomes study. *Medical Care* 43(4):391–400.

Koch CG, Li L, Hixson E, Tang A, Phillips S, Henderson JM. (2012) What are the real rates of postoperative complications: Elucidating inconsistencies between administrative and clinical data sources. *Journal of the American College of Surgeons* 214:798–805.

Kontopantelis E, Springate D, Reeves D, Ashcroft DM, Valderas JM, Doran T. (2014) Withdrawing performance indicators: Retrospective analysis of general practice performance under UK Quality and Outcomes Framework. *British Medical Journal* 348:g330.

Kosseim M, Mayo NE, Scott S, Hanley JA, Brophy J, Gagnon B, Pilote L. (2006) Ranking hospitals according to acute myocardial infarction mortality. Should transfers be included? *Medical Care* 44:664–670.

Kötter T, Blozik E, Scherer M. (2012) Methods for the guideline-based development of quality indicators – A systematic review. *Implementation Science* 7:21.

Krell RW, Finks JF, English WJ, Dimick JB. (2014) Profiling hospitals on bariatric surgery quality: Which outcomes are most reliable? *Journal of the American College of Surgeons* 219:725–734.

Krumholz HM, Lin Z, Drye EE, Desai MM, Han LF, Rapp MT, Mattera JA, Normand SL. (2011) An administrative claims measure suitable for profiling hospital performance based on 30-day all-cause readmission rates among patients with acute myocardial infarction. *Circulation: Cardiovascular Quality and Outcomes* 4:243–252.

Krumholz HM, Wang Y, Mattera JA, Wang Y, Han LF, Ingber MJ, Roman S, Normand SL. (2006) An administrative claims model suitable for profiling hospital performance based on 30-day mortality rates among patients with heart failure. *Circulation* 113:1693–1701.

Landon BE, Normand SL. (2008) Performance measurement in the small office practice: Challenges and potential solutions. *Annals of Internal Medicine* 148(5):353–357.

Landrum MB, Bronskill SE, Normand SLT. (2000) Analytic methods for constructing cross-sectional profiles of health care providers. *Health Services Outcomes Research Methodology* 1(1):23–47.

Langley GJ, Moen R, Nolan KM, Nolan TW, Norman CL, Provost LP. (2009) *The Improvement Guide: A Practical Approach to Enhancing Organizational Performance*, 2nd edn. San Francisco, CA: Jossey-Bass.

Larsen K, Merlo J. (2005) Appropriate assessment of neighborhood effects on individual health: Integrating random and fixed effects in multilevel logistic regression. *American Journal of Epidemiology* 161(1):81–88.

Langley GJ, Nolan KM, Nolan TW, Norman CL, Provost LP. (1996) *The Improvement Guide: A Practical Approach to Enhancing Organizational Performance*. San Francisco, CA: Jossey-Bass.

Lau D, Padwal RS, Majumdar SR, Pederson JL, Belga S, Kahlon S, Fradette M, Boyko D, McAlister FA. (2015) Patient-reported discharge readiness and 30-day risk of readmission or death: A prospective cohort study. *American Journal of Medicine* 129(1): 89–95.

Laverty AA, Laudicella M, Smith PC, Millett C. (2015) Impact of "high-profile" public reporting on utilization and quality of maternity care in England: A difference-in-difference analysis. *Journal of Health Services Research and Policy* 20(2):100–108.

Laverty AA, Smith PC, Pape UJ, Mears A, Wachter RM, Millett C. (2012) High-profile investigations into hospital safety problems in England did not prompt patients to switch providers. *Health Affairs (Millwood)* 31(3):593–601.

Lee RI, Jones LW. (1933) *The Fundamentals of Good Medical Care*. Chicago, IL: University of Chicago Press.

Lee SH, Cabana MD. (2006) Indices for continuity of care: A systematic review of the literature. *Medical Care Research Reviews* 63(2):158–188.

Lee SJ, Walter LC. (2011) Quality indicators for older adults preventing unintended harms. *Journal of the American Medical Association* 306(13):1481–1482.

Lee TH, Mongan JJ. (2012) *Chaos and Organization in Health Care*. Cambridge, MA: MIT Press.

Leslie M, Paradis E, Gropper MA, Reeves S, Kitto S. (2014) Applying ethnography to the study of context in healthcare quality and safety. *British Medical Journal Quality and Safety* 23:99–105.

Lim KG, Patel AM, Naessens JM, Li JT, Volcheck GW, Wagie AE, Enders FB, Beebe TJ. (2008) Flunking asthma? When HEDIS takes the ACT. *American Journal of Managed Care* 14(8):487–494.

Liu Y, Traskin M, Lorch SA, George EI, Small D. (2015) Ensemble of trees approaches to risk adjustment for evaluating a hospital's performance. *Health Care Management Science* 18:58–66.

Lohr KN (ed.) (1990) *Medicare: A Strategy for Quality Assurance*, Vols. I and II. Washington, DC: National Academy Press.

Lowry CA, Woodall WH, Champ CW, Rigdon SE. (1992) A Multivariate Exponentially Weighted Moving Average Control Chart. *Technometrics* 34(1):46–53.

MaCurdy T, Kerwin J, Gibbs J, Lin E, Cotterman C, O'Brien-Strain M, Theobald N. (2008) Evaluating the functionality of the Symmetry ETG and Medstat MEG software in forming episodes of care using Medicare data. Acumen, LLC, Burlingame, CA. Available at https://www.cms.gov/Research-Statistics-Data-and-Systems/Statistics-Trends-and-Reports/Reports/downloads/macurdy.pdf (accessed September 2015).

Mainz J. (2003a) Developing clinical indicators. *International Journal for Quality in Health Care* 15 (Suppl 1):i5–i11.

Mainz J. (2003b) Defining and classifying clinical indicators for quality improvement. *International Journal for Quality in Health Care* 15(6):523–530.

Mantel N. (1970) Why stepdown procedures in variable selection. *Technometrics* 12(3):621–625.

Marang-van de Mheen PJ, Shojania KG. (2014) Simpson's paradox: How performance measurement can fail even with perfect risk adjustment. *British Medical Journal Quality and Safety* 23:701–705.

Marshall C, Best N, Bottle A, Aylin P. (2004a) Statistical issues in the prospective monitoring of health outcomes across multiple units. *Journal of the Royal Statistical Society Series A* 167(3):541–559.

Marshall EC, Spiegelhalter DJ. (1998) Reliability of league tables of in vitro fertilisation clinics: Retrospective analysis of live birth rates. *British Medical Journal* 316:1701–1705.

Marshall M, McLoughlin V. (2010) How do patients use information on health providers? *British Medical Journal* 341:c5272.

Marshall MN, Shekelle PG, Leatherman S, Brook RH. (2000) The public release of performance data: What do we expect to gain? A review of the evidence. *Journal of the American Medical Association* 283:1866–1874.

Marshall T, Mohammed MA, Rouse A. (2004b) A randomized controlled trial of league tables and control charts as aids to health service decision-making. *International Journal for Quality in Health Care* 16(4):309–315.

Martin LA, Nelson EC, Lloyd RC, Nolan TW. (2007) *Whole System Measures. IHI Innovation Series white paper.* Cambridge, MA: Institute for Healthcare Improvement.

Martuzzi M, Hills M. (1995) Estimating the degree of heterogeneity between event rates using likelihood. *American Journal of Epidemiology* 141(4):369–374.

Mason W. (2001) Multilevel methods of statistical analysis. In: Smelser NJ, Baltes PB (eds), *International Encyclopedia of the Social and Behavioral Sciences.* Amsterdam, the Netherlands: Elsevier Science.

Matheny ME, Resnic FS, Arora N, Ohno-Machado L. (2007) Effects of SVM parameter optimization on discrimination and calibration for post-procedural PCI mortality. *Journal of Biomedical Informatics* 40(6):688–697.

Mathers CD, Lopez AD, Murray CJL. (2006) The burden of disease and mortality by condition: Data, methods and results for 2001. In: Lopez AD, Mathers CD, Ezzati M, Murray CJL, Jamison DT (eds), *Global Burden of Disease and Risk Factors.* New York: Oxford University Press, pp. 45–240.

Maybin J, Addicott R, Dixon A, Storey J. (2011) *Accountability in the NHS: Implications of the Government's Reform Programme.* London, UK: The King's Fund.

Mayer EK, Bottle A, Rao C, Darzi AW, Athanasiou T. (2009) Funnel plots and their emerging application in surgery. *Annals of Surgery* 249(3):376–383.

McClellan M, McNeil BJ, Newhouse JP. (1994) Does more intensive treatment of acute myocardial infarction in the elderly reduce mortality? Analysis using instrumental variables. *Journal of the American Medical Association* 272(11):859–866.

McConchie S, Shepheard J, Waters S, McMillan AJ, Sundararajan V. (2009) The AusPSIs: The Australian version of the Agency of Healthcare Research and Quality patient safety indicators. *Australian Health Review* 33(2):334–341.

McCormick N, Lacaille D, Bhole V, Avina-Zubieta JA. (2014) Validity of myocardial infarction diagnoses in administrative databases: A systematic review. *PLOS ONE* 9(3): e92286.

McCormick TH, Raftery AE, Madigan D, Burd RS. (2012) Dynamic logistic regression and dynamic model averaging for binary classification. *Biometrics* 68:23–30.

McDonald JH. (2014) *Handbook of Biological Statistics*, 3rd edn. Baltimore, MD: Sparky House Publishing.

Mehrotra A, Hussey PS, Milstein A, Hibbard JH. (2012) Consumers' and providers' responses to public cost reports, and how to raise the likelihood of achieving desired results. *Health Affairs (Millwood)* 31(4):843–851.

Meltzer DO, Chung JW. (2014) The population value of quality indicator reporting: A framework for prioritizing health care performance measures. *Health Affairs* 33(1):132–139.

Milch CE, Salem DN, Pauker SG, Lundquist TG, Kumar S, Chen J. (2006) Voluntary electronic reporting of medical errors and adverse events. *Journal of General Internal Medicine* 21:165–170.

Mittlböck M, Heinzl H. (2001) A note on R^2 measures for Poisson and logistic regression models when both models are applicable. *Journal of Clinical Epidemiology* 54:99–103.

Mittlböck M, Schemper M. (1996) Explained variation for logistic regression. *Statistics in Medicine* 15:1987–1997.

Mohammed MA, Manktelow BN, Hofer TP. (2012) Comparison of four methods for deriving hospital standardised mortality ratios from a single hierarchical logistic regression model. *Statistical Methods in Medical Research*.

Montgomery DC. (2012) *Introduction to Statistical Process Control*, 7th edn. London, UK: John Wiley & Sons.

Morris S, Hunter RM, Ramsay AIG, Boaden R, McKevitt C, Perry C, Pursani N et al. (2014) Impact of centralising acute stroke services in English metropolitan areas on mortality and length of hospital stay: Difference-in-differences analysis. *British Medical Journal* 349:g4757.

Moscoe E, Bor J, Bärnighausen T. (2015) Regression discontinuity designs are under-utilized in medicine, epidemiology, and public health: A review of current and best practice. *Journal of Clinical Epidemiology* 68:132–143.

Mugford M, Banfield P, O'Hanlon M. (1991) Effects of feedback of information on clinical practice: A review. *British Medical Journal* 303:398–402.

Mumford V, Greenfield D, Hinchcliff R, Moldovan M, Forde K, Westbrook JI, Braithwaite J. (2013) Economic evaluation of Australian acute care accreditation (ACCREDIT-CBA (Acute)): Study protocol for a mixed method research project. *British Medical Journal* Open 3:e002381.

Murray CJL, Frenk J. (2000) A WHO Framework for Health System Performance Assessment. *Bulletin of the World Health Organization* 78(6): 717–731.

Nardo M, Saisana M, Saltelli A, Tarantolo S, Hoffman A, Giovanini E. (2005) *Handbook on Constructing Composite Indicators: Methodology and User Guide*. Paris, France: OECD Publishing.

National Quality Forum. (2009) *Measurement Framework: Evaluating Efficiency Across Patient-Focused Episodes of Care*. Washington, DC: NQF.

National Quality Forum. (2012) *Multiple Chronic Conditions Measurement Framework*. Washington, DC: NQF.

Navarro V. (2000) Assessment of the world health report 2000. *Lancet* 356:1598–1601.

Needham DM, Scales DC, Laupacis A, Pronovost PJ. (2005) A systematic review of the Charlson comorbidity index using Canadian administrative databases: A perspective on risk adjustment in critical care research. *Journal of Critical Care* 20:12–19.

Neuhauser D. (2002) Heroes and martyrs of quality and safety: Ernest Amory Codman MD. *Quality and Safety in Health Care* 11:104–105.

Nicholl J, Jacques RM, Campbell MJ. (2013) Direct risk standardisation: A new method for comparing casemix adjusted event rates using complex models. *BMC Medical Research Methodology* 13:133.

Nightingale F. (1858) *Notes on Matters Affecting the Health, Efficiency, and Hospital Administration of the British Army*. London, UK: Harrison. Available at https://archive.org/details/b20387118 (accessed September 2015).

Nolan TP, Berwick DM. (2006) All-or-none measurement raises the bar on performance. *Journal of the American Medical Association* 295:1168–1170.

Normand SL, Glickman ME, Gatsonis CA. (1997) Statistical methods for profiling providers of medical care: Issues and applications. *Journal of the American Statistical Association* 92:803–814.

Normand SL, Wolf RE, Ayanian JZ, McNeil BJ. (2007) Assessing the accuracy of hospital clinical performance measures. *Medical Decision Making* 27(1):9–20.

O'Brien SM, Shahian DM, DeLong ER, Normand S-LT, Edwards FH, Ferraris VA, Haan CK et al. (2007) Quality measurement in adult cardiac surgery: Part 2—Statistical considerations in composite measure scoring and provider rating. *Annals of Thoracic Surgery* 83:S13–S26.

O'Donnell CA. (2000) Variation in GP referral rates: What can we learn from the literature? *Family Practice* 17(6):462–471.

OECD. (2014) *Geographic Variations in Health Care. What Do We Know and What Can Be Done to Improve Health System Performance?* Paris, France: OECD Publishing. Available at http://www.oecd-ilibrary.org/social-issues-migration-health/geographic-variationsin-health-care_9789264216594-en (accessed 9 December 2014).

OPM Evaluation Team. (2009) Evaluation of the healthcare commission's healthcare associated infections inspection programme. OPM report, London, UK, pp. 1–23.

Ovretveit J. (2002) Improving the quality of health services in developing countries: Lessons for the West. *Quality and Safety in Health Care* 11:301–302.

Padman R, McColley SA, Miller DP, Konstan MW, Morgan WJ, Schechter MS, Ren CL, Wagener JS; Investigators and Coordinators of the Epidemiologic Study of Cystic Fibrosis. (2007) Infant care patterns at Epidemiologic Study of Cystic Fibrosis sites that achieve superior childhood lung function. *Pediatrics* 119:E531–E537.

Palmer RH, Reilly MC. (1979) Individual and institutional variables which may serve as indicators of quality of medical care. *Medical Care* 17:693–717.

Parker JP, Li Z, Damberg CL, Danielsen B, Carlisle DM. (2006) Administrative versus clinical data for coronary artery bypass graft surgery report cards. The view from California. *Medical Care* 44:687–695.

Parkin D, Hollingsworth B. (1997) Measuring production efficiency of acute hospitals in Scotland, 1991–94: Validity issues in data envelopment analysis. *Applied Economics* 29(11):1425–1433.

Patten SB, Williams JV, Lavorato DH, Bulloch AG, D'Arcy C, Streiner DL. (2011) Recall of recent and more remote depressive episodes in a prospective cohort study. *Social Psychiatry and Psychiatric Epidemiology* 47(5):691–696.

Peabody JW, Luck J, Jain S, Bertenthal D, Glassman P. (2004) Assessing the accuracy of administrative data in health information systems. *Medical Care* 42:1066–1072.

Pepe MS, Janes H, Longton G, Leisenring W, Newcomb P. (2004) Limitations of the odds ratio in gauging the performance of a diagnostic, prognostic, or screening marker. *American Journal of Epidemiology* 159:882–890.

Perla RJ, Hohmann SF, Annis K. (2013) Whole-patient measure of safety: Using administrative data to assess the probability of highly undesirable events during hospitalization. *Journal of Healthcare Quality* 35(5):20–31.

Peters E, Dieckmann N, Dixon A, Hibbard JH, Mertz CK. (2007) Less is more in presenting quality information to consumers. *Medical Care Research and Review* 64:169–190.

Peterson ED, DeLong ER, Jollis JG, Muhlbaier LH, Mark DB. (1998) The effects of New York's bypass surgery provider profiling on access to care and patient outcomes in the elderly. *Journal of the American College of Cardiology* 32(4):993–999.

Peterson ED, Delong ER, Masoudi FA, O'Brien SM, Peterson PN, Rumsfeld JS, Shahian DM et al. (2010) ACCF/AHA 2010 position statement on composite measures for healthcare performance assessment: A report of the American College of Cardiology Foundation/American Heart Association Task Force on Performance Measures. *Circulation* 121:1780–1791.

Polancich S, Restrepo E, Prosser J. (2006) Cautious use of administrative data for decubitus ulcer outcome reporting. *American Journal of Medical Quality* 21:262–268.

Poloniecki J. (1998) Half of all doctors are below average. *British Medical Journal* 316:1734–1736.

Pope GC, Kautter J, Ellis RP, Ash AS, Ayanian JZ, Iezzoni LI, Ingber MJ, Levy JM, Robst J. (2004) Risk adjustment of Medicare capitation payments using the CMS-HCC model. *Health Care Financing Reviews* 25(4):119–141.

Porter M. (2010) What is value in health care? *New England Journal of Medicine* 363:2477–2481.

Potts HWW, Anderson JE, Colligan L, Leach P, Davis S, Berman J. (2014) Assessing the validity of prospective hazard analysis methods: A comparison of two techniques. *BMC Health Services Research* 14:41.

Preen DB, Holman CD, Spilsbury K, Semmens JB, Brameld KJ. (2006) Length of comorbidity lookback period affected regression model performance of administrative health data. *Journal of Clinical Epidemiology* 59(9):940–946.

Pringle M, Wilson T, Grol R. (2002) Measuring "goodness" in individuals and health care systems. *British Medical Journal* 325:704–707.

Profit J, Typpo KV, Hysong SJ, Woodard LD, Kallen MA, Petersen LA. (2010) Improving benchmarking by using an explicit framework for the development of composite indicators: An example using pediatric quality of care. *Implementation Science* 5:13.

Quality of Care and Outcomes Research in CVD and Stroke Working Groups. (2000) Measuring and improving quality of care: A report from the American Heart Association/American College of Cardiology First Scientific Forum on assessment of healthcare quality in cardiovascular disease and stroke. *Circulation* 1(12):1483–1493.

Quan H, Drösler S, Sundararajan V, Wen E, Burnand B, Couris CM, Halfon P et al. for the IMECCHI Investigators. (2005) Adaptation of AHRQ patient safety indicators for use in ICD-10 administrative data by an international consortium. *Advances in Patient Safety* 2(1).

Quan H, Parsons GA, Ghali WA. (2002) Validity of information on comorbidity derived from ICD-9-CCM administrative data. *Medical Care* 40(8):675–685.

Raftery J, Roderick P, Stevens A. (2005) Potential use of routine databases in health technology assessment. *Health Technology Assessment* 9(20):1–92.

Raleigh VS. (2012) Pick'n' mix: An introduction to choosing and using indicators. King's Fund, London, UK. Available at http://www.kingsfund.org.uk/sites/files/kf/field/field_document/Outcomes-choosing-and-using-indicators-the-king's-fund-aug-20121.pdf.

Reeves D, Campbell SM, Adams J, Shekelle PG, Kontopantelis E, Roland M. (2007) Combining multiple indicators of clinical quality an evaluation of different analytic approaches. *Medical Care* 45:489–496.

Render ML, Kim M, Deddens J, Sivaganesin S, Welsh DE, Bickel K, Freyberg R et al. (2005) Variation in outcomes in Veterans Affairs intensive care units with a computerized severity measure. *Critical Care Medicine* 33:930–939.

Research Triangle International. (2011) Centre for Medicare & Medicaid Services RTI Evaluation of FY 2011 Data, Chart A: Research Triangle International. Available at http://www.rti.org/reports/cms/ (accessed 2 September 2015).

Rosko MD, Mutter RL. (2008) Stochastic frontier analysis of hospital inefficiency. A review of empirical issues and an assessment of robustness. *Medical Care Research Reviews* 65(2):131–166.

Rosen R. (1991) *Life Itself: A Comprehensive Inquiry into Nature, Origin, and Fabrication of Life.* New York: Columbia University Press.

Rost N, Bottle A, Lee JM, Randall M, Middleton S, Shaw L, Thijs V, Rinkel GJE, Hemmen TM. (2016) Stroke severity is a crucial predictor of outcome: An international prospective validation study. *Journal of the American Heart Association.* 5(1):e002433.

Royal College of Nursing. (2009) Measuring for quality in health and social care. An RCN position statement. Publication no: 003 535. Available at https://www.rcn.org.uk/__data/assets/pdf_file/0004/248872/003535.pdf (accessed September 2015).

Royal College of Physicians. (2005) *Doctors in Society: Medical Professionalism in a Changing World. 2005.* Report of a working party of the Royal College of Physicians of London. London, UK: RCP.

Royston P, Moons KG, Altman DG, Vergouwe Y. (2009) Prognosis and prognostic research: Developing a prognostic model. *British Medical Journal* 338:b604.

Rubin DB. (1987) *Multiple Imputation for Nonresponse in Surveys.* New York: Wiley & Sons.

Rubin HA, Pronovost P, Diette G. (2001) The advantages and disadvantages of process-based measures of health care quality. *International Journal of Quality in Health Care* 13(6):469–474.

Runger GC, Alt FB, Montgomery DC. (1996) Contributors to a multivariate statistical process control signal. *Communications in Statistics—Theory and Methods* 25(10):2203–2213.

Ryan AM, Tompkins CP. (2014) *Efficiency and Value in Healthcare: Linking Cost and Quality Measures.* Washington, DC: National Quality Forum.

Saary MJ. (2008) Radar plots: A useful way for presenting multivariate health care data. *Journal of Clinical Epidemiology* 61(4):311–317.

Salmon JW, Heavens J, Lombard C, Tavrow P. (2003) *The Impact of Accreditation on the Quality of Hospital Care: KwaZulu-Natal Province, Republic of South Africa.* Operations Research Results, Vol. 2(17). Bethesda, MD: U.S. Agency for International Development (USAID), Quality Assurance Project, University Research Co., LLC, pp. 1–49.

Salomon JA, Murray CJL. (2002) The epidemiologic transition revisited: Compositional models for causes of death by age and sex. *Population and Development Review* 28(2):205–228.

Sanagou M, Wolfe R, Forbes A, Reid CM. (2012) Hospital-level associations with 30-day patient mortality after cardiac surgery: A tutorial on the application and interpretation of marginal and multilevel logistic regression. *BMC Medical Research Methodology* 12:28.

Sari AB, Sheldon TA, Cracknell A, Turnbull A, Dobson Y, Grant C, Gray W, Richardson A. (2007) Extent, nature and consequences of adverse events: Results of a retrospective casenote review in a large NHS hospital. *Quality and Safety in Health Care* 16(6):434–439.

Sexton JB, Helmreich RL, Neilands TB, Rowan K, Vella K, Boyden J, Roberts PR, Thomas EJ. (2006) The safety attitudes questionnaire: Psychometric properties, benchmarking data, and emerging research. *BMC Health Services Research* 6:44.

Shadish WR, Cook TD, Campbell D. (2002) *Experimental and Quasi-Experimental Designs for Generalized Causal Inference*. Boston, MA: Houghton Mifflin.

Shahian DM, Torchiana DF, Shemin RJ, Rawn JD, Normand SL. (2005) Massachusetts cardiac surgery report card: Implications of statistical methodology. *Annals of Thoracic Surgery* 80:2106–2113.

Sharabiani MT, Aylin P, Bottle A. (2012) Systematic review of comorbidity indices for administrative data. *Medical Care* 50(12):1109–1118.

Shekelle PG, Lim YW, Mattke S, Damberg C. (2008) *Does Public Release of Performance Results Improve Quality of Care? A Systematic Review*. London, UK: Health Foundation.

Shekelle PG, Woolf SH, Eccles M, Grimshaw J. (1999) Developing guidelines. *British Medical Journal* 318:593.

Sherman H, Castro G, Fletcher M, Hatlie M, Hibbert P, Jakob R, Koss R et al. (2009) World Alliance for Patient Safety Drafting Group. Towards an international classification for patient safety: The conceptual framework. *International Journal of Quality in Health Care* 21:2–8.

Shroyer AL, Grover FL, Edwards FH. (1998) 1995 coronary artery bypass risk model: The society of thoracic surgeons adult cardiac national database. *Annals of Thoracic Surgery* 65(3):879–884.

Shtatland ES, Kleinman K, Cain EM. (2002) *One More Time about R^2 Measures of Fit in Logistic Regression*. NESUG. Cary, NC: SAS Institute Inc.

Shwartz M, Ash AS, Anderson J, Iezzoni LI, Payne SM, Restuccia JD. (1994) Small area variations in hospitalization rates: How much you see depends on how you look. *Medical Care* 32(3):189–201.

Simborg DW. (1981) DRG creep: A new hospital-acquired disease. *New England Journal of Medicine* 304(26):1602–1604.

Smith PC, Mossialos E, Papanicolas I, Leatherman S (eds). (2009) *Performance Measurement for Health System Improvement: Experiences, Challenges and Prospects*. Cambridge, UK: Cambridge University Press.

Snijders TAB, Bosker RJ. (1999) *Multilevel Analysis: An Introduction to Basic and Advanced Multilevel Modelling*. London, UK: Sage Publications.

Spiegel A. (1997) Hammurabi's Managed Health Care – Circa 1700 B.C. Managed Care, May. Available at http://www.managedcaremag.com/archives/1997/5/hammurabis-managed-health-care-circa-1700-bc (accessed on 14 April 2016).

Spiegelhalter DJ. (2005) Handling over-dispersion of performance indicators. *Quality and Safety in Health Care* 14:347–351.

Stafoggia M, Lallo A, Fusco D, Barone AP, D'Ovidio M, Sorge C, Perucci CA. (2011) Spie charts, target plots, and radar plots for displaying comparative outcomes of health care. *Journal of Clinical Epidemiology* 64(7):770–8.

Staiger DO, Dimick JB, Baser O, Fan Z, Birkmeyer JD. (2009) Empirically derived composite measures of surgical performance. *Medical Care* 47:226–233.

Steiner SH, Cook RJ, Farewell VT. (1999) Monitoring paired binary surgical outcomes using cumulative sum charts. *Statistics in Medicine* 18:69–86.

Steiner SH, Cook RJ, Farewell VT, Treasure T. (2000) Monitoring surgical performance using risk-adjusted cumulative sum charts. *Biostatistics* 1(4):441–452.

Steyerberg EW, Eijkemans MJ, Harrell FE Jr, Habbema JD. (2000) Prognostic modelling with logistic regression analysis: A comparison of selection and estimation methods in small data sets. *Statistics in Medicine* 19(8):1059–1079.

Stukenborg GJ, Kilbridge KL, Wagner DP, Harrell FE Jr, Oliver MN, Lyman JA, Einbinder JS, Connors AF Jr. (2005) Present-at-admission diagnoses improve mortality risk adjustment and allow more accurate assessment of the relationship between volume of lung cancer operations and mortality risk. *Surgery* 138:498–507.

Sullivan LM, Massaro JM, D'Agostino RB Sr. (2004) Presentation of multivariate data for clinical use: The Framingham Study risk score functions. *Statistics in Medicine* 23:1631–1660.

Sundararajan V, Romano PS, Quan H, Burnand B, Drösler SE, Brien S, Pincus HA, Ghali WA. (2015) Capturing diagnosis-timing in ICD-coded hospital data: Recommendations from the WHO ICD-11 topic advisory group on quality and safety. *International Journal of Quality in Health Care* 27(4):328–333.

Sutherland K, Leatherman S. (2006) *Regulation and Quality Improvement: A Review of the Evidence*. London, UK: The Health Foundation.

Tan-Torres Edejer T, Baltussen R, Adam T, Hutubessy R, Acharya A, Evans DB, Murray CJL. (2003) *Making Choices in Health: WHO Guide to Cost-Effectiveness Analysis*. Geneva, Switzerland: World Health Organization.

Tate R, Wongbundhit Y. (1983) Random versus non-random coefficient models for multilevel analysis. *Journal of Educational Statistics* 8:103–120.

Teixeira-Pinto A, Normand SL. (2008) Statistical methodology for classifying units on the basis of multiple-related measures. *Statistics in Medicine* 27:1329–1350.

Testik MC, Borror CM. (2004) Design strategies for the multivariate EWMA control chart. *Quality and Reliability Engineering International* 20:571–577.

Thomas N, Longford NT. (1994) Empirical Bayes methods for estimating hospital-specific mortality rates. *Statistics in Medicine* 13:889–903.

Tibshirani R. (1997) The Lasso method for variable selection in the Cox model. *Statistics in Medicine* 16(4):385–395.

Timbie JW, Normand SL. (2008) A comparison of methods for combining quality and efficiency performance measures: Profiling the value of hospital care following acute myocardial infarction. *Statistics in Medicine* 27(9):1351–1370.

Tu JV, Donovan LR, Douglas SL, Wang JT, Austin PC, Alter DA, Ko DT. (2009) Effectiveness of public report cards for improving the quality of cardiac care. The EFFECT study: A randomized trial. *Journal of the American Medical Association* 302(21):2330–2337.

Turner T, Misso M, Harris C, Green S. (2008) Development of evidence-based clinical practice guidelines (CPGs): Comparing approaches. *Implementation Science* 3:45.

UK Biobank. (2007) UK Biobank Protocol 2013. Available at www.ukbiobank.ac.uk/resources/ (accessed September 2015).

U.S. Department of Veterans Affairs. Million Veteran Program. Available at http://www.research.va.gov/mvp/ (accessed September 2015).

Utter GH, Cuny J, Sama P, Silver MR, Zrelak PA, Baron R, Drösler S, Romano PS. (2010) Detection of postoperative respiratory failure: How predictive is the agency for healthcare research and quality's patient safety indicator? *Journal of the American College of Surgeons* 211:347–354.

Utter GH, Zrelak PA, Baron R, Tancredi DJ, Sadeghi B, Geppert JJ, Romano PS. (2009) Positive predictive value of the AHRQ accidental puncture or laceration patient safety indicator. *Annals of Surgery* 250:1041–1045.

Varabyova Y, Schreyögg J. (2013) International comparisons of the technical efficiency of the hospital sector: Panel data analysis of OECD countries using parametric and non-parametric approaches. *Health Policy* 112:70–79.

Veloski J, Boex JR, Grasberger MJ, Evans A, Wolfson DB. (2006) Systematic review of the literature on assessment, feedback and physicians' clinical performance: BEME Guide No. 7. *Medical Teacher* 28(2):117–128.

Vincent C. (2010) *Patient Safety*, 2nd edn. Chichester, UK: John Wiley & Sons.

Vincent C, Aylin P, Franklin BD, Holmes A, Iskander S, Jacklin A, Moorthy K. (2008) Is health care getting safer? *British Medical Journal* 337:a2426.

Vincent C, Burnett S, Carthey J. (2013) The measurement and monitoring of safety: Drawing together academic evidence and practical experience to produce a framework for safety measurement and monitoring. The Health Foundation. Available at www.health.org.uk/publications/themeasurement-and-monitoring-of-safety (accessed September 2015).

Viviani L, Zolin A, Mehta A, Olesen HV. (2014) The European cystic fibrosis society patient registry: Valuable lessons learned on how to sustain a disease registry. *Orphanet Journal of Rare Diseases* 9:81.

Wachter RM, Pronovost PJ. (2006) The 100,000 Lives Campaign: A scientific and policy review. *The Joint Commission Journal on Quality and Patient Safety* 32(11):621–627.

Wagner AK, Soumerai SB, Zhang F, Ross-Degnan D. (2002) Segmented regression analysis of interrupted time series studies in medication use research. *Journal of Clinical Pharmacy and Therapeutics* 27(4):299–309.

Walker K, Neuburger J, Groene O, Cromwell DA, van der Meulen J. (2013) Public reporting of surgeon outcomes: Low numbers of procedures lead to false complacency. *Lancet* 382(9905):1674–1677.

Wang D, Zhang W, Bakhai A. (2004) Comparison of Bayesian model averaging and stepwise methods for model selection in logistic regression. *Statistics in Medicine* 23(22):3451–3467.

Welke KF, Karamlou T, Diggs BS. (2008) Databases for assessing the outcomes of the treatment of patients with congenital and paediatric cardiac disease – A comparison of administrative and clinical data. *Cardiology in the Young* 18(Suppl 2):137–144.

Wennberg JE, Freeman JL, Culp WL. (1987) Are hospital services rationed in New Haven or over-utilised in Boston? *Lancet* 329(8543):1185–1189.

Williams SC, Schmaltz SP, Morton DJ, Koss RG, Loeb JM. (2005) Quality of care in U.S. hospitals as reflected by standardized measures, 2002–2004. *New England Journal of Medicine* 353:255–264.

Wilson RM, Runciman WB, Gibberd RW, Harrison BT, Newby L, Hamilton JD. (1995) The quality in Australian health care study. *Medical Journal of Australia* 163(9):458–471.

World Health Organization. (2011) Causes of death 2008: Data sources and methods. World Health Organization. (2014) World health statistics 2014. Available at http://apps.who.int/iris/bitstream/10665/112738/1/9789240692671_eng.pdf (accessed September 2015).

WHO, Department of Health Statistics and Informatics. (2011) Available at http://www.who.int/healthinfo/global_burden_disease/cod_2008_sources_methods.pdf?ua=1 (accessed October 2015).

Yeh RW, Vasaiwala S, Forman DE, Silbaugh TS, Zelevinski K, Lovett A, Normand SLT, Mauri L. (2014) Instrumental variable analysis to compare effectiveness of stents in the extremely elderly. *Circulation: Cardiovascular Quality and Outcomes* 7:118–124.

Zeng L, Zhou S. (2011) A Bayesian approach to risk-adjusted outcome monitoring in healthcare. *Statistics in Medicine* 30:3431–3446.

Zhan C, Miller MR. (2003) Administrative data based patient safety research: A critical review. *Quality and Safety in Health Care* 12(Suppl II):ii58–ii63.

Zuckerman IH, Lee E, Wutoh AK, Xue Z, Stuart B. (2006) Application of regression-discontinuity analysis in pharmaceutical health services research. *Health Services Research* 41(2):550–563.

Index

A

Accreditation
  COHSASA accreditation standards, 18
  compliance, 18
  financial costs, 17, 19
  Joint Commission, 17
  non-participation, 16
  patient care process and outcomes, 17
  program participation, 16
  quality improvement initiatives, 19
  quality performance, 18
  regulation, 16
ACHS, *see* Australian Council on Healthcare Standards
Acute coronary syndrome (ACS), 178
Acute myocardial infarction (AMI), 14, 21, 25, 42, 78, 139, 160
Acute Physiology and Chronic Health Evaluation (APACHE II), 105
Agency for Healthcare Research and Quality (AHRQ), 15, 72, 74, 136, 180
Akaike information criterion (AIC), 114
AMI, *see* Acute myocardial infarction
Analysis of variance (ANOVA), 28
Antimicrobial Point Prevalence Survey, 23
Artificial neural networks (ANNs), 113
Australian Council for Safety and Quality in Health Care (ACSQHC), 10, 13
Australian Council on Healthcare Standards (ACHS), 17
Australian Health Service Safety, 13
Australian National Safety, 223
Autocorrelation, 210

B

Bayesian additive regression trees (BART), 113
Bayesian analysis, 156, 235
Bayesian model averaging (BMA), 111
Bristol Royal Infirmary Inquiry, 9, 12, 21
Bundaberg Base Hospital (BBH), 10

C

CABG, *see* Coronary artery bypass graft
Care Quality Commission, 13, 17, 140, 174
CCGs, *see* Clinical commissioning groups
Centers for Medicare & Medicaid Services (CMS), 14, 33, 99, 136, 156, 197
Charlson index, 99
Claims-clinical mortality model, 81–82
Clinical Classification Software (CCS), 72, 182
Clinical commissioning groups (CCGs), 200
Clostridium difficile, 20, 23
CMS, *see* Centers for Medicare & Medicaid Services
Commonwealth Fund, 177
Component indicators, weights and
  all-or-nothing scoring, 143
  clinical and policy prioriti, 140
  common options, 141
  efficiency frontier approach, 144
  face validity and stakeholder buy-in, 143
  general practitioners, 143
  iatrogenic pneumothoraxes and "never events," 142
  pay-for-performance schemes, 142
  Pearson's coefficient, 142
  simplest weighting scheme, 141
  stakeholders, 141
  varimax rotation, 143
Composite measures
  AHRQ's Patient Safety Indicator Composite, 146–147
  alternative approaches

"all-or-nothing" combination process, 149
  Bayesian framework, 150
  latent variable model, 149
  randomised controlled trials, 149
  trouble-free pathway, 149
  within-and between-patient correlations, 150
construction steps
  choosing healthcare units, 138
  indicators, 138–139
  metrics, 140
  missing data values, 138
  run descriptive analyses, 138–139
  run sensitivity analyses, 144–145
  scope and purpose, 138
  spider diagrams/radar plots, 145
  "spie" charts, 145
  weights and component indicators, 140–144
examples, 136–137
Leapfrog Group, 147–148
Consumer Assessment of Healthcare Providers and Systems (CAHPS), 15, 20, 87
Coronary artery bypass graft (CABG), 11, 80, 96, 106
Cost–benefit analysis (CBA), 222
Cost-effectiveness analysis (CEA), 220
Cost-utility analysis (CUA), 222
Council for Health Services Accreditation of Southern Africa (COHSASA) program, 18
Cross-national databases
  challenges
    clustering within countries, 184
    culture-specific sensibilities, 180
    International Database, 181
    "soft" outcomes, 180
    stroke mortality, 181–183
    unusual data/apparent performance, 184–185
  indicators, 177
  international comparisons, data sources, 185–186
  multinational patient-level databases, 178–180
  patient records, 177
CUA, see Cost-utility analysis

D

Danish Healthcare Quality Programme, The, 18
Danish National Indicator Project, 199
Dartmouth Atlas Project, 11
Data envelopment analysis (DEA), 57–58
Data sources
  administrative data
    coding systems, 71–74
    Farr's disease classification, 70
    ICD system, 70
    patient safety events, 74–75
    pattern identification and hypotheses formulation, 71
    statutory/financial purpose, 70
  administrative vs. clinical database accuracy
    accurate electronic record, 78
    CABG clinical risk model, 80
    casemix adjustment and hospital comparisons, 82
    Charlson score, 79
    claims model, 81–82
    comorbidities and complications, 77
    discrimination, 82
    Elixhauser comorbidity, 79
    indirect standardisation and random effects models, 81
    missing data, 77
    PPV, 78
    public reporting, 80
    SMRs, 80
  clinical registry data, 75–77
  clinical trials and registries, 90
  data quality, 67–69
  non-response bias, 89
  paper-based formal surveys, 90
  patient websites, 89
  secondary use of, 90
  surveys, 87–89

voluntary and mandatory reporting systems, safety events
   active surveillance, 84
   AHRQ, 86
   Minimal Information Model, 86
   NHS organisation risk management system, 85
   NRLS, 85
   SWIFT, 84
DEA, *see* Data envelopment analysis
Deductive reasoning, 223
Diagnosis-Related Group (DRG) system, 72
Dickey–Fuller statistic, 211
Difference-in-differences (DID) analysis, 214–216
Disability-adjusted life-years (DALYs), 53
Dr. Foster Intelligence, 12
Drug-eluting stent (DES), 217
Durbin–Watson statistic, 210

E

Effectiveness measures, 14
Electronic health records (EHRs), 76
Electronic medical record (EMR), 100
Elixhauser index, 99
Empirical Bayes (EB) estimate, 167
"End Results System," 8
Episode of care, 32
Episode Treatment Group® (ETG), 32
Escherichia coli infections, 13
European Cystic Fibrosis Society Patient Registry, 179–180
European Journal of Public Health, 180
European Statistics Code of Practice, 145

F

Failure mode and effect analysis (FMEA), 84
False alarm rate (FAR), 233
False discovery rate (FDR), 163
Freedom of Information Act, 11

G

Generalised estimating equations (GEEs), 129
Glasgow coma scale (GCS), 103

Global Burden of Disease study, 185
Global Comparators project, 179
Global Registry of Acute Events (GRACE), 178
Global Trigger Tool, 84
Greater Manchester model, 215
Gross domestic product (GDP), 186

H

Hammurabi's system, 7
Hand hygiene, 13
Healthcare-associated infections (HCAIs), 23
Healthcare Commission, 12–13
Health Care Financing Administration, 11
Healthcare performance data
   format
      accountability, 194
      barriers, effective feedback, 194–195
      evaluability, 193
      feedback, 193
      formulating strategy, 194
      healthy culture, organisation, 194
      value sensitivity, 192
   giving performance information, units
      audit and feedback, 202
      balanced scorecard, 198
      benchmarking and bespoke analytics, 202
      CCGs, 200
      Danish healthcare system, 200
      Gapminder World video plots, 202
      Hospital Compare, 204–205
      NHS choices and data screenshots, 203
      NHS Improving Quality, 199
      public-facing websites, 203
      Right Care Data Tools, 200
      scatter plot, 201
   managers and clinicians, 195–197
   metadata, 206
   presentation
      compare indicators, 191
      evaluative cues, 190
      fictitious unit-level data, 190

HEAT targets, 190
  radial plot, 191
  spider plot, 192
  "spie charts", 191
  target plots, 191
  traffic light display, fictitious data, 190
  types, 189
results presentation, public, 197–198
Healthcare Resource Group (HRG) system, 72
Health Plan Employer Data and Information Set (HEDIS), 14, 18
Healthy life expectancy (HALE), 53
HEDIS, see Health Plan Employer Data and Information Set
HES, see Hospital Episodes Statistics
Hibbard paper, 22
Hosmer–Lemeshow statistic, 114
Hospital Episodes Statistics (HES), 35, 78, 105
Hospital standardised mortality rate (HSMR), 101, 174
Hospital-wide Mortality Indicator, 35
Hotelling T^2 control chart, 165
Hub-and-spoke model, 215

I

ICC, see Intraclass correlation coefficient
Ideal quality indicators, 59, 60
IHI, see Institute for Healthcare Improvement
Imputation, 233–234
Indicators
  candidate indicator, 63–64
  desired health outcomes, 38
  efficiency and value
    cost-effectiveness, 53
    DEA, 57–58
    measured quality and patient health needs, 56
    NQF, 54
    SFA, 57–58
    total healthcare spending, 54
    "unnecessary" treatment, 53
  guidelines, 59
  ideal indicator, 59–60
  indicator taxonomies
    Care Quality Commission, 39, 41
    domains, 39
    equity, 41
    improved transparency and accountability, 42
    intermediate outcomes, 39
    never events, 41
    outcome indicators, 39
    patient-centredness echoes, 41
    performance measurement system, 42
    quality of care, 38
    structure and process indicators, 39
  multimorbidity, 45–47
  outcomes, 61
  patient safety
    alarming estimates, 42
    global movement, 42
    Health Foundation, 44
    incident reporting systems, 43
    inter-rater reliability, 43
    organisational culture, 44
    staff attitudes, 44
    working practices, 45
  population health and healthcare system quality
    avoidable mortality, 49
    efficiency indicators, 49
    equity, 49
    health action, 48
    international comparisons, 48
    prenatal monitoring, 49
    responsiveness, 49
    stakeholder consultation, 49
    WHO World Health Statistics report, 50–53
  risk adjustment, 61
  structure and process measures, 59
  validation of, 62
  withdraw indicators, 64–65
Inductive reasoning, 223
Institute for Healthcare Improvement (IHI), 152, 174
Instrumental variable (IV) analysis, 214, 216–217
Interactive voice response (IVR), 88
International Classification for Patient Safety (IC4PS), 43

Index

International Classification of Diseases (ICD), 71
International Consortium for Health Outcomes Measurement (ICHOM), 62
International Statistical Classification of Disease (ICD), 182
Interquartile range (IQR), 167
Interrupted time series (ITS), 210
Intraclass correlation coefficient (ICC), 139, 170, 231
Inverse probability of treatment weighting (IPTW), 214
Item response theory (IRT) models, 149

J

The Joint Commission, 14
Journal of the American Medical Association (JAMA), 76
Journal of the Statistical Society of London, 8

K

King Edward Memorial Hospital (KEMH), 1, 10

L

Leapfrog Group, 14
Low-and middle-income countries (LMICs), 17

M

Manchester Patient Safety Framework (MaPSaF), 44
Marshall's ranking method, 156
MCAR, *see* Missing completely at random
Median odds ratio (MOR), 170
Medical Care Availability and Reduction of Error (MCARE) Act, 87
Medicare Hospital Outcomes Survey (HOS) program, 14
Medicare Mortality Predictor System, 101
Medicare Provider Analysis and Review (MEDPAR), 81
Methicillin-resistant *Staphylococcus aureus* (MRSA) infections, 210
Mid Staffordshire NHS Hospitals Trust, 1
Missing completely at random (MCAR), 102
Model Hospital Statistical Form, 70
Modifiable areal unit problem (MAUP), 28
Modified Early Warning Score, 94
Modified modified continuity index (MMCI), 46
Monte Carlo simulation, 214
Multiple chronic conditions (MCCs), 45
Multiple testing, 234–235
Myocardial Ischaemia National Audit Project (MINAP), 78

N

National Cardiac Database (NCD), 11
National Cardiovascular Intelligence Network, 201
National Clinical Coding Qualification, 74
National Committee for Quality Assurance (NCQA), 14, 18
National Coordinating Council for Medication Error Reporting and Prevention, 43
National Healthcare Quality Reporting System, 63
National Health Performance Authority, 13
National Health Performance Framework, 49
National Health Service (NHS), 12, 175, 190, 209
National Institute for Health and Care Excellence (NICE), 53
National Institutes of Health Stroke Scale (NIHSS), 183
National Patient Safety Agency (NPSA), 85
National Quality Forum (NQF), 34, 45
National Reporting and Learning System (NRLS), 43, 85

National Safety and Quality Health Service Standards, 13
National Surgical Quality Improvement Program (NSQIP), 75, 110, 128
NCQA, *see* National Committee for Quality Assurance
New South Wales Bureau of Health Information, 13
Normand's patient profile method, 156
NRLS, *see* National Reporting and Learning System
NSQIP, *see* National Surgical Quality Improvement Program
Nursing Home Quality Initiative, 197

O

Optum™, 32
Organisation for Economic Cooperation and Development (OECD), 48, 63, 177

P

Patient-reported outcome measures (PROMs), 41, 149
Patient safety indicators (PSIs), 74–75, 136, 146
Pay-for-performance (P4P) scheme, 35, 64
Performance monitoring system
  aviation and manufacturing fields, 1
  continuous quality improvement, 3
  evaluation
    adaptive trial technique, 209
    CEA, 220–221
    confounders, adjusting approaches, 212–214
    DID, 214–216
    economic analysis basic types, 220
    explanatory trials, 208
    instrumental variable analysis, 216–217
    interrupted time-series design and analysis, 209–212
    natural experiments, 209
    quality improvement (QI) projects, 209
    RCT, 208
    regression discontinuity designs, 217–219
    "uncontrolled before and after comparison," 208
  false alarms, 3
  financial costs, healthcare systems, 1
  global patient safety movement, 1
  goals of
    accountability, 15–16
    informed consent, 23–24
    openness and transparency, 21
    patient choice, 19–20
    prevent harm and unsafe care, 22–23
    professionalism, 23
    quality improvement, 21–22
    regulation and accreditation, 16–19
  healthcare scandals
    Bristol Inquiry, 9
    casemix, 11
    Commission Inquiry, 11
    data quality and completeness, 10
    Douglas Inquiry, 10
    "high-risk mortality," 9
    incident monitoring and mortality review systems, 10
    Medicare data, 11
    national trends, 12
    New York's post-CABG mortality, 12, 24
    patient selection, 12
    responses, 12–13
    risk-adjusted cardiac surgical outcomes, 11
    Shipman analysis, 12–13
    statistical techniques, 9
    "switch" cardiac procedure, 9
  measuring and monitoring quality, 2
  methodologists, 3
  monitoring schemes, 14–15
  origins, 7–8
  risk-stratification and decision support tools, 227
  simple *vs.* complex, 225
  specific *vs.* general, 226–227
  trade-offs, 226

Index 267

Performance threshold
  acceptable variation, 167–170
  Bayesian methods, providers comparison, 155–157
  historical benchmarks, 153
  inter-unit variation, 153–155
  multiple testing
    back-of-the-envelope calculation, 164
    Benjamini–Hochberg procedure, 164
    Bonferroni correction, 163
    false alarm rate, 163
    FDR, 164
    multivariate statistical process control methods, 165–166
    SPC, 166
  outlier status, fixed vs. random effects, 171–172
  quality improvement methods, 174–175
  reliability, 172–174
  SPC and funnel plots, 157–162
  targets, 152–153
  variation between units, 166–167
Picker Institute, 41
Poisson model, 115
Positive predictive value (PPV), 68, 74, 169
Professional Association of Clinical Coders website, 74
Propensity score matching (PSM), 213
PSIs, see Patient safety indicators
Public Health England HCAI, 23

Q

Quality Accreditation Scheme, 13
Quality-adjusted life-years (QALYs), 63, 222
Quality and Outcomes Framework (QOF), 153
Quality Health Service, 223

R

RAND Corporation, 30, 63
Random effects model
  average patient, 128
  'average quality' hospital, 128
  empirical Bayes estimates/shrunken estimates, 127
  fixed hospital effects, 126
  poor performers, 128
  "smoothed" mortality, 128
  statistical preferences, 127
Random intercept model, 156
Randomised controlled trials (RCTs), 26, 178, 208
Receiver operating characteristic (ROC) curve, 114
Regression discontinuity (RD) designs, 214, 217–219
Relative risk (RR), 122, 123, 126, 128
Risk-adjusted outcomes
  Bayesian methods, 130
  CMS, 132
  confidence intervals, 131
  externally validated risk adjuster, 130
  fixed effects approach, 126
  hierarchical models, 131
  marginal vs. multilevel models, 129–130
  NSQIP, 132
  random effects, SMRs, 126–129
  ratios vs. differences, 122
  "shrunken" estimates and multilevel models, 131
  standardisation and logistic regression, 123–126
  volume–outcome relation, 132
Risk-adjustment models
  artificial neural network, 232
  calibration, 231
  clustering, 231
  fixed effects, 231
  logistic regression model, 230
  machine learning method, 232
  multilevel/hierarchical models, 230
  random effects, 230–231
  regression, 229
  shrinkage, 230
  standardisation, 229
Risk-adjustment principles and methods
  age and socioeconomic status, 98
  alternatives, 96
  candidate variables, 101–102

comorbidity, 98–100
continuous variables, 111
contraindications, 95
data, model fitting of
   adjusted R^2, 115
   Bayesian paradigm, 118
   discrimination/calibration, 117
   "fudge factor," 117
   Hosmer–Lemeshow statistic for calibration, 116–117
   regression coefficients into risk score conversion, 118–119
   ROC curve, 115–116
   socioeconomic deprivation and inverse care law, 118
disease severity, 100–101
factors, 96–98
final set of variables
   BMA, 111
   clinical judgment, 110
   cross-validation, 109
   mean squared error, 109
   misclassification error rate, 109
   shrinkage techniques and hybrid techniques, 110
   "stepwise" procedures, 110
   validation subset, 108
missing and extreme values, 102–104
patient preferences, 95
risk factor measurement, timing of, 104–107
and risk prediction, 94–95
statistical method for modelling
   ANNs, 113
   BART, 114
   hierarchical modelling, 111
   ICC, 112
   "machine learning" techniques, 112, 114
   regression methods, 111
   shrunken estimates, 112
structure and process measures, 95
Risk stratification, 96

S

Seasonality, 210
Society of Thoracic Surgeons (STS), 11
Spain's Atlas, 179
SPC, *see* Statistical process control
Spiegelhalter's ranking method, 156
Standardised mortality ratios (SMRs), 122, 153, 156
   aggregation levels, 124
   Byar's approximation, 126
   confidence interval, 125
   direct standardisation, 125
   indirect standardisation, 123
   logistic regression model, 123
   Poisson distribution, 125
   relative risk, 123
   Simpson's paradox, 124
   types of, 133
Staphylococcus aureus, 13, 23
Statistical process control (SPC), 153
   alarm limits, 233
   common cause variation, 232
   control limits, 158, 232
   cumulative sum, 233
   cumulative sums, 159, 161
   EWMA, 160, 233
   false alarm rate, 161
   FAR, 233
   funnel plot, 160
   funnel plots, 232
   IHI and The Joint Commission, 157
   in-control chart, 161
   outlying performance, 162
   out-of-control chart, 162
   overdispersion, 233
   run chart, 232
   run charts, 157–158
   SDR, 233
   Shewhart charts/control charts, 157, 159
   special cause variation, 232
   "special cause" variation, 159
   unusual performance, 159
   VLAD plots, 160
   warning limits, 233
Stochastic frontier analysis (SFA), 57–58
Structural equation modelling (SEM), 144
"Structured what-if technique" (SWIFT), 84
Successful detection rate (SDR), 233
Summary hospital-level mortality indicator (SHMI), 174

Summary Hospital Mortality
    Indicator, 101
Support vector machines (SVMs), 113
Symmetry®, 32
Systematic component variation
    (SCV), 167

T

Test–retest reliability, 69

U

UK Data Protection Act, 90
Unit of analysis
    clustering
        aggregation, 30
        ANOVA, 28
        "cluster-robust" standard
            errors, 27
        disease incidence rates, 31
        ECHO project, 31
        frequent monitoring, 27
        geographical area, 31
        intraclass correlation, 26
        marginal/population-averaged
            models, 28
        methodological issues, 29
        multilevel (aka hierarchical)
            modelling, 28
        population-averaged model, 28
        randomisation, 26
        space and time, 30
    episode treatment groups, 32–34
Unit of reporting, 34–36

U.S. Dartmouth Atlas of Health
    Care, 179
U.S. Society of Thoracic Surgeons
    Quality Measurement Task
    Force, 141
Usual provider of care (UPC), 46

V

Ventilator-associated pneumonia
    (VAP), 152

W

Winsorisation, 104
World Health Organization (WHO), 43,
    48, 70, 152, 177
    Minimal Information Model, 86
    MONICA database, 184
    Patient Safety Programme, 175
    World Health Statistics report
        all-cause mortality, 52
        DALYs, 53
        global health indicators, 51
        goals, 50
        HALE, 53
        health-related Millennium
            Development Goals, 51
        Spain assessment, 50
        YLL, 52–53

Y

Years of life lost (YLL), 52–53
Yellow Card Scheme, 85